PRACTICAL HINTS ON INFRA-RED SPECTROMETRY FROM A FORENSIC ANALYST

PRACTICAL HINTS ON
INFRA-RED SPECTROMETRY
FROM A FORENSIC ANALYST

M. J. de Faubert Maunder

B.Sc., A.R.I.C. F.R.A.S.

Deputy Training Officer, Laboratory of the Government Chemist
Assistant Secretary of the Chilterns and Middlesex Section of the Royal Institute of Chemistry

ADAM HILGER
LONDON

ISBN 0 85274 122 7

Published by

ADAM HILGER LTD
Rank Precision Industries Ltd
31 Camden Road, London NW 1

PRINTED BY BELL AND BAIN LTD, GLASGOW

PREFACE

I wish to record my grateful thanks to those who have been directly responsible for this book. The majority are now anonymous in the mists of time, but their contribution, outside the field of infra-red work, to my education and approach to life has been substantial. One such teacher stands out in my memory, and I must mention him by name—Professor Parkinson, whose books should, I feel, be on every analyst's bookshelf.

Several people have been intimately involved in the preparation of this book, and I thank them most sincerely. Dr R. J. Mesley made many constructive comments during the writing, while Miss Joyce Sadler managed, somehow, to interpret and type an essential portion of the text when it had been written. The thought of producing a comprehensive index was becoming a nightmare and I am greatly indebted to Mr H. G. Baxter for taking on the enormous task. My brother Lawrence, the artistic member of the family, supplied the essential artwork from which the illustrations are derived. Finally, I record my thanks to my wife who made the whole exercise possible by tolerating the unholy muddle for months on end, and for giving so freely of her time.

M. J. de Faubert Maunder

Teddington
March 1971

CONTENTS

INTRODUCTION

> It is by being conversant with the inventions of others that we learn to invent, as by reading the thoughts of others we learn to think.
>
> JOSHUA REYNOLDS

Papers, articles and books on the theoretical aspects of infra-red spectroscopy abound, and details of band assignments or correlations are printed in similar profusion. Few authors have paid more than token lip service to the intricacies of sample preparation and its attendant problems, and fewer still have given concrete examples of the topics they have discussed, before plunging into the theory behind spectroscopy. The basic theory is well taught at schools and colleges, and all workers must by now be familiar with this and can correctly assign major bands to a particular functional group. A further book on this topic would therefore be presumptuous.

A modern family car, like a commercial spectrometer, has a minimum of essential controls, but before a man is considered competent to drive it unsupervised he must undergo a lengthy period of detailed expert tuition and pass a driving test to prove his ability. Before any reliance can be placed on his results, the operator of a spectrometer must be satisfied that he is competent to 'drive' a machine, and recognize his own shortcomings or limitations.

Practical tests of ability are virtually non-existent beyond college level, and even then they only indicate a minimum standard of competence. One of the few genuine tests of chemical practical ability is provided by the Royal Institute of Chemistry's examination for the Mastership of Chemical Analysis, and in this,

A*

the amount of spectroscopy is, of necessity, restricted. Unless a person specializes in infra-red spectroscopy at some stage of his career, he will never take the opportunity to delve into the finer points and gain the necessary expertise.

It takes very little longer to complete sample preparation correctly than in some inadequate manner. At a time when 'job satisfaction' is one of the 'in' phrases, there is an aesthetic pleasure in being able to complete a task under adverse conditions, whereas a fixed routine (by 'numbers') can be soul destroying to the extent that a person becomes a liability through sub-standard work. A challenge, or variety, is the spice of life. The few attempts that have been made to give detailed practical instructions have been restricted by a shortage of examples and have largely been ignored in a climate hostile to manual skill; it is one of the strange anachronisms of a technical age, the inverse respect paid to craftsmanship. This book is an unashamed attempt to bring back an element of craftsmanship and critical appraisal of results. Because a forensic analyst must have the confidence to defend his expert opinion against critical cross-examination in the Courts, he must set himself impeccable analytical standards, and this book is offered to all infra-red spectrometrists in the hope that they will emulate his example. Courts are at liberty to demand analytical evidence, and if some of the infra-red spectra submitted are shown to be faulty technically, reasonable doubt can be thrown on the analyst's competence in any aspect of his work, particularly if he has promulgated his errors via the open literature. This book is, therefore, written with the forensic scientist in mind, for, at all times, he must produce work accurate 'beyond all reasonable doubt'. A fallacious result in normal circumstances will have no more serious an outcome than a disproved theory or, at worst, a ruined reputation, but in forensic work an individual's freedom, or even life, may be at stake. However, although the book takes its illustrations mainly from the forensic field, its practical advice is addressed to all workers in the infra-red.

By bringing into the open some of the more glaring technical errors, it is hoped to raise the general standard of infra-red work as a whole, and emphasize the often ignored facet of the importance of the individual working a machine. How often has one heard the facile phrase, 'This machine is no good'?

Solution techniques are generally well taught and discussed and provide few problems, even in quantitative applications, and nearly all workers can return reliable results almost from the beginning of their working lives. Discussion has therefore been centred on the much more unpredictable solid sampling techniques and their problems. The nature of the sample is an important parameter and the degree of dispersion during preparation is highly dependent on it. In the absence of expert first-hand guidance, the subtle changes in approach can only be learnt the hard way. Some second-hand guidance is possible if the faults can be diagnosed from a standard set of illustrations of the type given in this book. These illustrations cannot be comprehensive by the very nature of the diversity of real-life samples and an overlap of faults, and it is confidently anticipated that this work will be found deficient in its turn. Until that time it may prove to be better than nothing.

Specific examples to illustrate points made in the text are chosen from my own experience of products regulated by the principal Drugs Acts in Great Britain. The penalties attendant on illicit possession can be as high as 14 years' imprisonment and it is therefore vital that no shadow of doubt shall remain in an identification. It is assumed that the reader is familiar with the fundamental chemical methods of isolating poisons from biological materials by all the techniques capable of presenting an extract suitable for infra-red examination. A short annotated bibliography is given to collate some of the more novel (mainly chromatographic) approaches.

Very little of the material written here is original but the method of presentation is intended to highlight facts and opinions which ought to be more widely known. Some points have been given more than once so that, for reference purposes, many chapters are self-contained. The references chosen to amplify the theme are necessarily only a selection which appear to have a direct relevance. A few others of possible value have been appended for perusal. The method of presenting references is intended to assist the reader in selecting further reading and should overcome some of the shortcomings of a more normal format where an occasional printing error renders the whole reference valueless; some of those cited here were culled from other sources and required some considerable research to discover the correct original. By giving the whole original title (with obvious abbreviations where appropriate), a quick assessment of the relative value is possible.

An essential adjunct to any analytical scheme is an efficient retrieval system. In infra-red spectrometry, the proposals and counter-proposals in the literature are not necessarily of direct application to an individual's needs, and a short review of the proposals with some comments on custom-built schemes is given.

To summarize, some hints and examples of the basic solid sampling techniques are reviewed as a necessary first step before a spectrum is accepted for interpretation, either practically or theoretically. Allied techniques and data retrieval are also discussed briefly. No claims are made for a comprehensive coverage, but if this book can assist some workers, or promote a less inhibited discussion of results, it will have served some useful purpose. Operators in other fields may find the itemized sample preparation data and fault diagnosis of some value since it is intended to follow the principles laid down by Miller in *Laboratory Methods in Infra-red Spectroscopy* (Sadtler Research Laboratories Inc., 1965) and by White in *Handbook of Industrial Infra-red Analysis* (Plenum Press, New York, 1964). My book also indirectly stresses the importance of self reliance and initiative and, it is hoped, a *soupçon* of the critical faculties, and will be found to have much in common with the approach of the books *Spectroscopic Tricks*, (edited by L. May, Adam Hilger Ltd, London; Plenum Press, New York, 1968) and *Practical Hints on Absorption Spectrometry* (*Ultra-violet and Visible*), (J. R. Edisbury, Adam Hilger Ltd, London; Plenum Press, New York, 1968).

I

SOLID SAMPLING

Error is but opinion in the making.
MILTON

PRESSED DISKS

In 1952, Schiedt and Reinwein in Germany, and Sisters Stimson and O'Donnell in America independently introduced the KBr disk technique.[173,183] The basic simplicity of the method, coupled with its ability to produce good spectra from very small samples, led to its very wide adoption. In spite of a number of practical difficulties noted over the years, the technique remains the best all round sampling method for solids where qualitative identification is required, and can be made quantitative with sufficient skill. Nearly all materials of interest are solids, either in their own right or as a salt produced in the isolation process, and the pressed-disk technique is a corner stone in their analysis by infra-red spectrometry. This is particularly true when trace analysis is required, since no other technique is as effective or as easy to manipulate.

Most authorities follow the recipe:

1. Grind sample and matrix to a particle size of 5 μm or smaller.
2. Mix sample at 0·5 per cent and grind/mix further as appropriate.
3. Transfer to an evacuable die and evacuate as highly as possible for at least 2 minutes.
4. Apply pressure (more than 100 kg/mm^2) for a minimum of 2 minutes.

5. Release pressure and vacuum and run spectrum as quickly as possible with a minimum of exposure to atmospheric moisture.

Technique is all important and may explain why some workers regard pressed disks with suspicion. Certainly, a considerable number of papers discuss aspects causing difficulties, but these should not cause a total disenchantment. It is convenient to discuss each aspect in the sequence in which it applies to the recipe above.

Matrix

KBr is frequently the material of choice, although how often this choice is due to a blind following of a published 'recipe' is a matter of conjecture. The important criteria are in order of importance:

1. High transmission of radiation (over waveband of interest).
2. Refractive index close to sample.
3. Low sintering pressure.
4. High purity and anhydrous.
5. Chemically stable.

These are the criteria of Butz[22] in a different order. Alkali halides comply with some or all of these and the choice largely rests on a suitable refractive index match. Schiedt[172] noted that KCl, KI and NaCl gave lower transmissions than KBr, but against this, KI and KCl are less prone to pick up atmospheric moisture. Hales and Kynaston[86] gave details of the preparation of a suitable grade of KCl, although Analytical Reagent (AR) grade is of adequate purity these days. KCl is easier to obtain in high purity, and at a lower cost, than KBr, and, being more brittle, is more readily ground by hand without forming a solid cake. Certainly there would appear to be an increasing use of KCl by many workers, and in pharmaceutical analysis where the majority of samples handled are hydrochloride salts, its use in preference to the bromide is strongly advocated for comparative purposes. Mismatches in refractive index are always possible and the reader should always be prepared to change the matrix.

A few other matrixes have been used in recent years. In 1959,

Teflon was tried because of its chemical inertness,[170] but it suffers from a lack of transparency and has strong absorption between windows. For the investigation of esters, amines and benzene substitution in the region 1600 to 1300 cm^{-1}, this matrix cannot be used because of total absorption, and would appear to be little used now except for specialist studies. Polythene is being used more often as a matrix in its own right, or as thin films in cheap and simple disposable cells. It is finding its greatest value as a matrix for investigating bands in the far infra-red (below 600 cm^{-1}), but as the majority of cheap instruments available to the average worker will not cover this region, it has limited application in micro-analysis.

The only non alkali halide matrix likely to find any widespread use is AgCl. Chen and Gould[33] describe a micro technique using this material and Spittler and Jaselskis[178] outline a simple method of manufacture. As AgCl is totally anhydrous, moisture absorption is no problem, and because it is so soft, disks of surprising clarity can be produced with deceptive ease. To some extent, these plastic properties of AgCl are an advantage and good spectra can be produced from opaque disks, presumably because the disk has some of the properties of a mull. Provided a simple method of micronizing or powdering AgCl can be adopted, this material should have a promising future.

Matrix Pretreatment

For most qualitative work it is not necessary to prepare a special grade of matrix other than grinding and drying commercial AR grade chemicals. This is particularly true if the same batch is used as a reference blank in a double-beam spectrometer. Some manufacturer's material will be better than others with respect to interference from nitrate or other impurities, and it will pay to experiment, purchasing bulk quantities when a satisfactory batch is encountered, rather than laboriously purifying a batch.

Most workers develop their technique, or lack of it, relatively quickly, and the choice of pretreatment adopted tends to be based on personal preference. Provided a relatively uniform particle sized powdered commercial product is available, good results can be obtained without pretreatment, using the material as received. Prolonged exposure of a finely ground (and meticulously prepared) matrix to the atmosphere during sample preparation is just as likely to pick up as much moisture as a relatively crude commercial product, stored in bulk, and prepared *in situ* with the sample. It is difficult to match the moisture content of a control precisely, particularly if the sample being examined is moist, contains water of crystallization in variable amounts, or is hygroscopic. Some imbalance is almost inevitable in these cases and it pays to consider the return on the effort involved in preparing a highly dried matrix.

Commercial alkali halides are not always finely powdered and some preparation is needed, even if this is only a crude crushing and sieving for size. Most workers agree on a mesh range between 200 and 400 at this stage. If the apparatus is available, the best product is obtained by freeze drying of an aqueous solution, although precipitation can be almost as effective. Hales and Kynaston[86] were advocates of precipitation, and obtained fine crystals by pouring saturated KCl solution into ice-cold concentrated HCl acid. The high cost of HBr and HI, and the solubility of their potassium salts, deters most people from trying the technique on these materials, although Kopff *et al.*[102] were able to eliminate interference from both water and nitrate in KBr by a subsequent drying at elevated temperatures.

Most workers will only have a hand pestle and mortar available for crushing, and will find this more than adequate. A good quality one is advisable to avoid localized 'fusion' or 'dead' spots, and with it an experienced worker will be able to complete a small batch of sample as rapidly as with a ball mill. Except for large batches, a ball mill is often more inconvenient because of the subsequent cleaning time, and unless the size of balls used, and the speed setting, are chosen with care, grinding is poor, the ultimate being a solid cake in the vessel, or an excessive proportion of 'fines'.

After sieving for size, 2 hours drying at 110°C removes all moisture for most practical purposes; the product can be stored in a closed vessel or in a desiccator over fused $CaCl_2$.

AgCl requires special treatment before it is suitable as a matrix. Spittler and Jaselskis made their starting material with 2*M* solutions of $AgNO_3$ and HCl.[178] Dark-room working and copious washing to remove residual nitrate are essential. Moisture was removed in a vacuum oven. In practice it is not necessary to filter off the precipitate or use a vacuum oven as the precipitate is sufficiently dense and coarse grained to tolerate many washings and decantations without loss, and drying can be completed in the original beaker in an ordinary oven at 110°C for one day.

Whether AgCl is manufactured as described or purchased in lump or sheet form, it is necessary to grind it before use. Because of its softness, this is virtually impossible at room temperature and Spittler and Jaselskis advocate cooling and grinding under liquid nitrogen. The reader may find it more convenient to experiment with different manufacturing techniques involving precipitation, subsequent drying in thin layers (to avoid caking), and different grinding methods to incorporate the sample, treating this stage more like a mull. Acceptable qualitative spectra can be obtained from what appears, to those familiar with KBr as a matrix, an unpromising batch of prepared AgCl.

Whatever method is adopted for matrix preparation, it is important that the method can be followed by any member of the analytical team using the spectrometer, and be reproducible on this basis over a number of years. To get full benefit from past spectra in an archive, this point cannot be overemphasized; the simpler the method consistent with the results demanded the better. Milkey[139] notes that overgrinding can produce adverse results and this is a warning that over enthusiasm can rebound. He ground small batches as needed in a plastic vessel. If the reader has sufficient confidence to wish to try more elaborate preparations, he is referred to the work of Von Dietrich,[195] Finkel'shtein *et al.*[77] or Ingebrigtson and Smith.[94] It is worth noting that Ingebrigtson and Smith found that Harshaw's optical grade KBr, consisting of random-sized crystals, was satisfactory, and this was confirmed by Kirkland,[100] indicating that crystal-size composition is as important as purity.

Accurate weighing of the matrix on a repetitive basis is not only tedious but unnecessary even in most quantitative work. A 2 per cent accuracy for the spectrometer is the best normally expected and consequently a 5–10 mg error on 500 mg matrix is permissible, thus allowing the use of one of the rapid action, direct reading balances. Some form of volume measuring device is a simpler alternative and in its crudest form will be a pile of powder on the end of a spatula which is judged by eye and experience. Mitzner[144] gave details of a simple measuring device based on different sized holes in a metal block. The wide range of gelatine capsules now available for drug presentation can be pressed into service as convenient measuring scoops. A dab of old-fashioned fish glue will bond the capsule cap or body to a wood dowelling handle at any convenient angle and will dry to a hard glossy surface. A laboratory inevitably collects a wide range of capsules in the course of its work and a virtually inexhaustible supply of measuring scoops is assured; selection from bodies as well as caps will cover all volumes of immediate value, or they can be trimmed to cover any specific intermediate volume. Used capsules are easily cleaned by a combination of a small camel-hair brush, air blasts and rinsing with powdered matrix (batches of matrix failing to meet a mesh range criterion are an obvious economy here). Mitzner used capsules in a similar manner and as a convenient method of storage.[143]

Sample Preparation

More contention is caused by the diversity of techniques applied at the sample preparation stage than by almost any other aspect of solid sampling. The object of the exercise is to disperse the sample uniformly in the matrix and in such a manner that an optimum spectrum is obtained. As with all manual skills, opinions will always clash over the best way of achieving this objective, and in an attempt to standardize conditions, the majority of workers advocate some form of mechanical grinding and/or mixing and achieve excellent qualitative and, in many cases, quantitative results. This approach can cause difficulties if it is taken to be a panacea in all cases—samples will not always behave in the same manner and a mechanical device can only carry out a pre-determined programme built into it. A worker approaching the subject for the first time can easily produce inferior results by following a published 'recipe' too blindly, a pitfall inherent in transmitting what is essentially a manual skill by print rather than by example. Slightly different tolerances in machine design coupled with minor variations in sample weight can have an amazing effect on grinding efficiency.

In circumstances where an authentic reference standard is available and can be given the same treatment as an unknown, the sample preparation technique adopted is of little consequence for qualitative identification, provided the conditions are similar for both. Good technique comes into its own when quantitative results are required or when authentic specimens are not available. The time spent in developing techniques capable of reliably reproducing spectra at least as good as those in the literature is time well spent; published spectra are the result of some experience on the part of the author, and it should not be forgotten that there is a natural human tendency to choose only the best spectra for publication.

It is generally agreed that a technique involving mixing of a pre-ground sample, rather than grinding with the matrix, gives better results. Kirkland[100] recommends this, especially for the preparation of samples difficult to grind. His results compare favourably with liquid phase measurements. Grinding a sample with the matrix may cause excessive fragmentation of the matrix, with adverse effects; the most common effect is moisture absorption. For achieving the necessary degree of grinding and mixing, a Wig-L-Bug amalgamator is invariably specified. A recent addition to the British market is the McCrone micronizing mill which works on a slightly different principle (Plate 1). It shows promise of giving every satisfaction once the appropriate speed, timing and loading conditions for the grinding jar have been established.

A danger inherent in all mechanical grinders/mixers is the introduction of interference from the machine itself. McDevitt and Baun[129] discussed contamination of KBr disks from plastic mixing vials, giving examples. Reduction of the mixing time, or increasing the amount of matrix, alleviated, but did not eliminate, the problem, and they concluded that a better solution would be to find an inexpensive vial which would not be abraded. A subsequent suggestion by Polchlopek and Robertson[154] of using acrylic plastic balls in place of the normal steel balls is not entirely satisfactory and does not eliminate all interference. McCrone reduced interference by specifying an agate grinding cylinder.

To summarize, it is essential to select carefully the correct material for the grinding vessel.

Other mixing methods finding wide application involve the use of a solvent at some stage. If the sample is soluble in water, a simple solution of both can be evaporated to dryness, ground if necessary, and pressed into a disk. Control of moisture content is difficult with hygroscopic samples, but this difficulty would arise anyway with a prolonged dry grinding, and can be alleviated by a 'wet' control blank in the reference beam. The ultimate in perfection is claimed for a lyophilized product, arising out of the work of Scheidt.[172,173] Where the apparatus and time are available, there can be no doubt about the efficiency of this concurrent mixing and dispersion of small particles. Sample volatilization can be quite serious, however. Organic samples insoluble in water can be dispersed by solvents; an addition of a solution to a powdered matrix is a crude technique, but may suffice in quick monitoring exercises where finesse is not required. Better results are obtained by grinding the matrix under a solution of the sample in some volatile solvent. In the early days of halide disks, Ingebrigston and Smith[94] recorded enormous increases in resolution and intensity, even when the sample was only slightly soluble in the solvent. This version is widely practised.

When dealing with small samples of the order of 1 mg or less, mechanical grinding is somewhat fatuous. The majority of workers will already possess a small agate pestle and mortar, and in many cases this will be all the apparatus available. A skilled worker familiar with the problems with mulls will have no difficulty in preparing small samples for halide disk examination, the accent being on skill (see p. 24). Applying too much pressure to small samples produces a solid cake in the mortar which is not readily removed by subsequent grinding with the matrix, and may simply aggravate the situation by producing an undesirable proportion of fines in the matrix. The grinding action should be just sufficient to cause an even mixing without excessive grinding, and may be impossible because of the poor design of the majority of mortars and pestles on the market—two examples from the same firm frequently have totally distinct grinding properties and can only be selected by experience. Where sample caking has occurred, the situation can be recovered somewhat by the addition of a relatively coarse matrix to act as an abrasive, and grinding until a requisite particle size is obtained. A compromise between the two extremes is possible by adding small portions of matrix and triturating these before attempting a final mixing which can be by mechanical means if precise quantitative work is desired. More extreme dispersion methods involve ultrasonics and heating and have not found widespread use.

There is one other method of sample dispersion which has not received the publicity it deserves. Shortly after the introduction of disk techniques, Clauson-Kaas and his co-workers[39] noted the difficulty of getting consistent results and solved the problem by a two-stage process: Make a disk in the normal manner, regrind and repress before presenting it to the spectrometer. It is surprising that this technique has not been more widely adopted as it is standard practice in the manufacture of pharmaceutical tablets of 'difficult' materials. Some compounds will not cohere adequately in a tablet on one run through a press, but the crude granules made in this run, reground as needed, have the necessary irregular shape and particle size (controlled by the second grinding stage) to cohere into a good tablet on the second run. Homogeneity is also better, and is the object of the present exercise. For both grindings, the particle size can be kept relatively large, thus avoiding excessive moisture pick-up, and is also within the capability of workers without mechanical grinding aids, eliminating a large element of personal grinding technique. In certain circumstances, an ordinary non-evacuable press will suffice, presumably owing to a more even crystal size and distribution, with a lower moisture content.

Pressing Disks and Pellets

In choosing a press, thought should be given to future work and a press of ample capacity selected capable of dealing with this. The extra costs involved will be repaid by lower maintenance costs on a machine working well within its capability for most of its working life, and most modern presses are now sufficiently robust to cater for all but the most extreme conditions.

Because of the wide variety of commercial presses available and the range of samples and matrices investigated, there is little agreement on the best conditions. When the conditions specified by a number of workers are analysed, 13 mm disks can be produced in 2 to 5 minutes at loads of 5000 kg or less for an average thickness of up to 5 mm. Larger sizes take proportionally higher loads and/or times, a load of over 10 000 kg maintained for 20 minutes being not unusual for a 25 mm diameter disk. A high vacuum should be maintained for a period before and during pressing, the precise conditions being determined by experiment. If possible, the vacuum should be derived from a backing pump for ease of handling and its relative 'hardness'.

In most work the size of sample is often severely limited and a need for pressing micro disks or pellets becomes apparent quickly. The majority of spectrometers have a relatively small beam size where the sample is placed, and in most spectrometers the beam, at this point, is rectangular in shape since it is essentially a facsimile of the objective slit. It is worth investigating the variation in size of this rectangle over the wavelength range of the instrument, particularly at long wavelengths when the objective slit is opened to its maximum. Provided a die can be made to approximate to the maximum size rectangle found, there is little point in pressing a pellet bigger than this and small samples can be concentrated where they are needed, in the beam. There are many different micro and semi-micro pellet masks available commercially and with the high sensitivity of modern spectrometers, spectra can be brought on-scale by quite large attenuations of the reference beam, normally achieved by placing a similar micro pellet mask, with a blank pellet in the reference beam. Commercial die blanks are designed to fit into a normal 13 or 25 mm die in a press and are convenient in use, but for maximum efficiency it may be found necessary to make one's own from cardboard or similar material. Edwards gave details of a neat die needing no evacuation.[70]

Using a die mask is simplicity itself. The prepared sample is placed in the mask in a normal die, and pressed. Lower pressures are needed than for a full-sized disk as the micro pellets are usually considerably thinner, of the order of 1 mm. The main advantage of using semi-micro pellet masks for all samples is the relative ease of handling and storage.

Plate 2 illustrates a commercial ultra micro KBr die in which, for the smaller pellets, the only pressure applied is provided by the atmosphere acting on the die.

Somewhere in between the standard sized and micro presses lies the Wilks Mini-Press. For low cost and ease of application nothing can really approach it. The press consists of a stainless steel barrel, serving subsequently as a pellet holder, with a $\frac{3}{8}$ inch × 18 thread* tapped through it, and two hexagon stainless steel bolts with polished ends acting as dies. One of the bolts is screwed into the barrel a fixed distance (5 turns), 50 to 100 mg prepared sample placed in the cavity, levelled by tapping the barrel, and the pellet pressed by screwing the second bolt home with a 9/16 × 6 inch spanner. After about 1 minute the two bolts are removed and the barrel containing the pellet is presented to the instrument with a special mounting supplied with the press. The pellet so formed is presented in a position close to the focus of most instruments, and in spite of a ring of opaque material around the central clear disk, transmissions of the order of 60 per cent are easily achieved. An evacuable version of the Mini-Press is available.

The Mini-Press will tolerate an amazing amount of abuse and quite opaque pellets with transmissions of 10 per cent or less can be attenuated in the reference beam to give sufficient detail for a qualitative identification.

Another practical advantage of the Mini-Press over the conventional press is its ability to reproduce light scattering in the reference pellet. Where an opaque sample pellet is produced and

* American thread form not Whitworth, but Whitworth will give a 'loose' fit.

cannot be improved, the reference pellet can be tailored to reproduce this opacity and energy scatter by a careful control of the pressure applied to the bolts, each try involving an increased pressure on the same pellet and offered-up to the spectrometer until the best fit is achieved.

Common Faults

The literature supplied with the Wilks Mini-Press itemizes some of the pitfalls encountered and how to overcome them, and it is relevant to discuss them here as they have general application in pressing disks.

1. *Pellet scatters energy at short wavelengths.* This is characterized by a sloping base-line with improving transmission towards the longer wavelengths; the pellet may not be obviously opaque on cursory inspection. The Christiansen filter effect may be noticed (this effect is discussed on pp. 66 and 146). Further grinding of the sample and/or matrix is required and is usually achieved in practice by punching out the pellet, regrinding and repressing.

2. *Isolated opaque spots in the pellet.* Isolated opaque spots are similar to 1 in cause and are due to uneven grinding of either the sample or matrix. The effect on the spectrum may pass unnoticed in mild cases.

3. *Cracked or flaked pellet.* Excessive grinding prior to pressing can cause the pellet to crack or flake. Wilks' literature suggests grinding fresh material but this may not be possible with unique samples, and the situation is usually redeemed as in 1 by regrinding and repressing. If the cracking is not excessive, acceptable spectra can be obtained without remedy, though with a lower overall transmission due to internal reflections, and occasionally interference pattern effects may be noticed.

4. *Opaque pellet over all or part of the area.* Overall transmission is characteristically low and the spectrum may exhibit Christiansen filter effects. The circumstances should indicate which of three likely causes is reponsible—inadequate pressure, bad distribution,

or the nature of the sample. Too high a concentration can normally be discounted.

(*a*) *Inadequate pressure* is cured by reassembling and repressing in most cases. Apart from the obvious cause of lack of applied pressure, excessive sample charge is the most common cause, although inadequate cleaning from a previous sample can leave an embedded crust in the thread of the barrel which prevents the die screws from exerting their full pressure over the whole area. This latter situation is quite common where faults 1 and 3 above have occurred and the offending pellet has been punched out for regrinding. With unique samples there is little option but to return the sample to the press after regrinding, where, with the slight but inevitable loss involved, the hindering effect of the residue in the thread can be appreciable. Provided the transference losses are minimal, firmer application of the *same* bolts in the same positions top and bottom should have the desired result, although it is undoubtedly preferable to avoid leaving an excessive amount of material in the threads in the first place. Careful 'chasing' with a micro-spatula is effective, if somewhat tedious, and quicker results are obtained by using another bolt specially kept for this purpose, thus avoiding damage to the polished faces of the die bolts, or, better still, the correct plug tap (free of lubricant!).

Where opaque pellets are consistently obtained with the Mini-Press and can be traced to inadequate pressure, the purchase of better quality and/or longer spanners should be considered as it is a false economy to buy anything less than 6-inch chrome vanadium ones; dipped rings cause less wear. In severe cases, a different, and more robust, member of the staff should be employed at this stage, although this is rarely necessary if the sample is correctly prepared in the first place.

(*b*) *Bad distribution* is characterized by localized areas of opacity, particularly a visual 'one-sided' appearance. The spectrum usually will be truncated with major peaks attenuated and may be useless even for qualitative work because of the overemphasis of minor bands. Any attempt to make even a semi-quantitative estimation is useless and the reader is referred to the work of Koenig for a full discussion of the wedge effect.[101] Care taken in levelling the sample is time well spent. Careful tamping with a micro-spatula is normally adequate, and in cases where the die can be removed in conventional presses, gentle tapping, normally by dropping a number of times onto a solid surface, is simpler. Several levelling devices have been described, and of these, Morgan's is as simple as any: The flat end of a rod is grooved, lowered to the surface of the powder, rotated slowly and removed slowly whilst still rotating.[147]

(*c*) *Nature of the sample.* Without some experience, the effects of the nature of the sample are difficult to distinguish from those of inadequate pressure and are usually characterized by a very even opacity or translucence. Some surprisingly good spectra can be obtained by the simple expedient of attenuating the reference beam provided the spectrometer has been properly commissioned beforehand.

The main distinguishing feature of this condition is a relatively high infra-red transmission in spite of the visual opacity, and the sharpness and general overall quality of the spectrum, with the major bands having approximately the absorption expected for the particular concentration involved.

The nearest analogy is to describe the pellet as a solid mull where the concentration is too high to form a true solid solution but is sufficiently well dispersed to act as a mull does with a closely matched refractive index. No amount of repressing or regrinding will alleviate the condition as it is due to the relative incompatibility of the sample and matrix. Dilution with more matrix is one answer provided it is not carried to excess giving a weak spectrum. Changing the matrix is an alternative in some cases, but it should always be borne in mind that such a change may cause a fundamental change in the resulting spectrum, and where no authentic reference standard is available, direct comparison with published spectra may not be valid. As a general rule, KCl is much more tolerant than KBr and yields clear pellets at much higher concentrations than those recommended for KBr.

5. *Irregular 'blotchy' appearance of the pellet.* At first glance an irregular 'blotching' appearance seems to be a more extreme version of 2 easily mistaken for 4. It is only distinguished by the pro-

nounced water bands observed in the spectrum. The cure recommended is to dry and break-up the prepared matrix. This is unlikely to be the normal cause since batches of prepared matrix are usually dried before storage, and even in non-hermetically sealed bottles on an open bench over a period of several months rarely pick up excessive moisture. Bad technique in the sample preparation stage is a much more reasonable explanation. Provided the sample is known to be dry or without water of crystallization, this source can be discounted. Conversely, it can be allowed for, or taken at its face value as a characteristic of the sample. As described in the section on sample preparation (p. 10ff), excessive grinding, and in particular prolonged grinding of the matrix, gives a very fine material capable of absorbing a considerable amount of moisture from the atmosphere. This is common to all subsequent pressing techniques and is usually alleviated by evacuating the dies before and during the pressing operation or by placing a reagent blank in the reference beam, or both. One of the claims for the evacuable Mini-Press is its ability to remove moisture during the evacuation stage. In view of the excellent pellets obtained by many workers in the past using unevacuated dies, and the inability of others to duplicate their work without evacuation, one is driven inevitably to the conclusion that some element of personal idiosyncrasy is responsible over and above natural variations in laboratory humidity.

The average worker would never consider fingering the face of a die but nevertheless it is not uncommon, the premium placed on those with naturally moist hands being particularly severe. A much more logical explanation is the habit of some people, mainly those with short-sight, to inspect every stage of their work closely. One breath of exhaled air will undo many hours of patient work. People who naturally breath through their mouths and thereby direct exhaled air straight at the object of their attention, rather than away as through the nose, are anathema as far as infra-red work is concerned and need special tuition. Where circumstances demand, dry box working may be necessary as recommended by Price and Maurer for unstable samples.[157]

One variable of pellet preparation not discussed in this section, except in passing, is control over the quantities of material used. Unless some control is exercised over the concentration of sample in a matrix, and the amount of this mix taken for pressing, one or more of the pitfalls 1–5 may occur. A concentration of sample in KBr around 0·5 per cent is standard practice, but a 1 per cent mix is possible in a number of cases. If working to a published method, the concentration used is invariably stated and this is based on optimum experience for the most part. Another reason for working to a constant concentration for all samples is the extra specificity this gives in the overall intensity of the resulting spectrum. By the same token, the quantity taken for pressing can affect the pressing efficiency and the smallest practical amount is advisable, unless, of course, a press will only function on a fixed volume before pressing. An intrinsically weak spectrum is sometimes encountered and some accurate idea of the quantities involved is essential for future reference before increasing the sample concentration and/or thickness of pellet to improve the spectrum for unambiguous identification.

Pellet Holders

Each spectrometer manufacturer supplies a good range of suitable accessories. In a number of cases these are derived from a specialist firm and it may pay to negotiate direct. Provided the pellet is supported in a plane perpendicular to the incident radiation, any crude mounting will suffice and may be little more than a piece of cardboard with a hole in it. For a person starting infra-red work it is a false economy to make pellet holders at least until some experience has been gained in the capabilities of the commercial items, the majority of which are well made and leave little to be desired.

Micro-work with micro-pellets usually requires a special holder. If a proprietary press is used some recommendations on the type of holder will be available; few of them require a subsequent removal of the pellet (Plate 3). It is becoming more common to press the pellet into a cavity in a sheet of inert material which either serves as a pellet holder in its own right, or, being the same size and shape as a standard disk, is mounted in a conventional sized holder. In the absence of a variable attenuator for the reference beam, a duplicate holder should be used to recover a maximum benefit from the lower transmission through a micro-holder.

de Klein and Ulbert solve the problem of pellet removal by using a Teflon insert in their die.[60]

Storage

Unless special precautions are taken, storage of pressed disks is not very satisfactory. Moisture is the major contaminant picked up and can be avoided by storage in a desiccator, although for any appreciable number of samples this becomes cumbersome. The problem is solved to some extent by making each disk self-contained in its own air-tight storage vessel. Micro-pellets, whether mounted or not, are just as simply treated as full-sized disks.

One of the earliest suggestions, and still as good as any, is the method of McGaughran.[130] Cut thin-walled polythene tubing into 2-inch lengths and staple each in the middle to make two compartments. As each compartment is filled, staple the end (or use a paper clip as a temporary measure). Code each compartment (sample) with a number and attach this number with adhesive tape. A 'Magic Marker' is a more modern expedient. Store these flat pellet files in a desiccator or a drying cabinet. McGaughran notes that, for 13-mm round pellets (the normal size of blank holder for a micro-pellet), tubing 0·5 inch in diameter and with a 0·008 inch wall is convenient, and information on ten such samples can be filed on a 5 × 3 inch card. Any excess prepared sample could be stored in the other half of each container unit, in which case, cross contamination through any gaps in the stapled dividing wall is not a serious matter.

In recent years, clear re-sealable plastic bags have become available. Some of these have a white patch on them capable of taking ball-point-pen writing and are excellent for storage purposes. A wide range of sizes is available.* Correctly sealed, the bags are virtually air-tight and need little special treatment other than

* Available from the Supreme Novelty Company, Vale Road, London, N.4.

storage in a drying cabinet to allow for small amounts of moisture diffusion over a long period. Excess prepared sample can be stored in a small specimen tube within the same pack and, with sensible handling, it will not damage delicate pellets. The main advantages of this type of pack are the cheapness ($3 \times 3\frac{1}{2}$ inch bags cost less than 0.25p) and the convenience of having all relevant information on the pack itself without the tedium of preparing and affixing a separate label or card index. Damaged, recrushed or small pellets can be stored in small specimen tubes within one of these sealable bags, and need no special storage conditions because of the double sealing. This is a very convenient method of storage where it is the intention to repress a disk rather than attempt to retain the delicate original intact. Excess prepared sample is simply added to the same tube to account for any physical losses incurred in the process.

Changes in the spectrum must always be anticipated during storage, particularly if moisture exclusion is not very efficient. Such changes can be turned to good account as an extra diagnostic aid. Examples of spectral changes on storage in a cell are cited in Chapter 7, p. 156.

The sample need not be ionic to change significantly on storage with moisture present.[10]

MULLS

A mull is defined as a suspension of small particles in a chemically inert liquid, prepared by trituration. Correctly prepared, it is intermediate in physical properties between a liquid and a solid.

It has been convenient to describe pressed disks first because this technique is applicable both to a wider range of samples and a wider range of sample sizes than the mull technique. At the same time, the pressed-disk technique suffers from more potential pitfalls which are easier to describe this way round. In circumstances where large samples are always to hand and the whole spectrum is not needed (say, in quality control), a worker need never use anything other than a mull preparation in his whole working life. In the converse situation with micro-samples, or the necessity to study the whole spectrum (in particular, the 3500 to 3000 cm^{-1} region), or in assays, serious consideration of the pressed-disk technique becomes obligatory. Quantitative work is not simple with mulls. It needs very careful attention to detail, a factor defeating the essential simplicity of a mull.

Because of the mull technique's essential simplicity and speed, together with its ability to return a spectrum of some sort every time, liberties tend to be taken with it. One of the arguments against a mull is that the liquid used for making it interferes strongly. This impression has been obtained by analysts following methods which advocate more liquid than is necessary. At the same time there is a strong disinclination to grind samples manually when, with the pressed-disk method, all stages of the technique are usually carried out mechanically, which may not necessarily be the most efficient manner of proceeding.

Where there is ample sample available (more than 1 mg) and a simple identification only required, as in 'white powders' submitted for confirmation of other tests, a good mull is quicker and simpler to prepare than a mediocre pressed disk, and considerably quicker than a good pressed disk.

Liquid

All the early work was carried out with a mineral oil. One of those offered for sale was very suitable and, through wide usage, has tended to lend its trademark name of 'Nujol' to the technique. With improvements in specifications over the years, there is no reason why an ordinary Paraffin oil, B.P. should not be used in place of Nujol, and many workers do so with complete success. However, even when this is done, the result is, more often than not, still colloquially referred to as a 'Nujol mull'.

Virtually all published mull spectra still use paraffin oil as the suspending medium and allow a direct comparison between them. There is some element of tradition in this, but, more important, paraffin oil is readily available at low cost (although the quantities used are negligible), it has virtually no volatility in a heated laboratory, is non-toxic and is odourless. The spectral regions obscured by paraffin oil's own absorption are also regions of normally limited interest—in a correctly made mull, these bands will not be obtrusive anyway and some evidence for those other bands due to the sample should be seen. In cases where C-H bands are of value, a non-hydrocarbon medium is required. Of the many possibilities, only two find general usage—hexachlorobutadiene, and perfloro-hydrocarbons sold under the name of 'Fluorolube'. The latter is less volatile and more widely used. Halocarbon oil is claimed to possess no interfering absorption between 4000 and 1300 cm^{-1}.[47] Removal of halocarbons from cell windows is difficult and the risk of sample contamination in subsequent examinations is high.

Mull Preparation

1. *Grinding*

As in the preparation of a pressed disk, efficient reduction of particle size is essential, perhaps more so. Although mechanical grinding can be used, it is uncommon in practice, and simple hand grinding with an agate pestle and mortar is almost universal. Most powders submitted for analysis are in an obviously crystalline form; most extracts tend to have an amorphous texture, if for no other reason than inadequate material for the niceties of recrystallization. In either event, a 1 to 3 mg sample is ample for the purpose in hand and no more than this should be taken. Experience will tell how the sample is ground with respect to pressure and motion of the pestle; there is no substitute for experience. One stands about a fifty-fifty chance of possessing a pestle and mortar which are totally useless for this delicate work, either through mismatches in radii of curvature or undersized pestles. Unless the radii and size are matched, grinding will only take place at one localized 'high spot' which will not only tend to fuse the sample into an intractable smear, but promote polymorphic and other crystal changes. In view of the high cost of agate equipment, inadequate sets should be returned until a satisfactory set is received—the author possesses two supposedly identical sets from the same firm, each with vastly different characteristics, one of which is fit for scrap only (without re-lapping). As always happens in real life, this latter was received first and accepted because it was not recognized for what it was at the time, its real character only becoming apparent in the fullness of time with increasing experience. This situation would appear to be more common than is freely admitted.

Grinding is continued until the particle size is reduced sufficiently, and must be very thorough to achieve uniformity. This primary stage reduces crystals to a size where radiation scatter is minimal, and efficiency at this point is more important than subsequently, in other words, effort should be devoted towards making this stage the only grinding necessary. A minimum of 1 minute's continuous grinding is required for soft and amorphous solids and 2 to 3 minutes for hard crystals (e.g. rock salt hardness). 2 minutes is a long time! This chore cannot be skimped and patience is needed. A convenient 'rule-of-thumb' to apply as a guide on grinding times is to increase the time taken as the cube of sample size. Increasing sample charge from 3 to 5 mg does not need just a simple increase of $1\frac{2}{3}$ but nearer 4 to 5 times. It is at this point that major faults in technique occur because grinding is a manual skill, not a science.

Mechanical grinding, if desired, can be carried out with the same apparatus as that used for pressed disks. Lohr and Kaier[113] describe a micro-capsule for the Wig-L-Bug, which is suitable for 3 to 5 mg samples. Not everyone would be able to manufacture this to the requisite degree of accuracy, but where it is possible, the capsule is simple and convenient and effective in operation on most mechanical shakers. There are two widely used alternatives to mechanical and manual grinding in a pestle and mortar.

(*a*) Flat, ground-glass plates make an excellent grinding tool and can be used on any type of sample. Very hard samples can abrade the glass with two important results—firstly, extra peaks appear in the spectrum and, secondly, fine glass particles ultimately have a deleterious effect on NaCl disks without very careful handling, especially during the dismantling of cells after use.

There is an important variation to this approach which is used at the Royal Institute of Chemistry's Summer Schools.

Instead of flat glass plates a 'Quick-fit' joint and stopper are used as grinding tools. Pyrex glass is very tough and with it higher grinding pressures can be applied without abrasion of the glass. There are also the added advantages of ready availability and low cost.

(*b*) Very soft samples can be ground between disks ultimately to be used as the cell. A number of amine salts are too hard to be safely ground this way, but most free bases or acids are already soft and/or amorphous and ideally suited for this grinding *in situ*.

Sticky materials, if they cannot be spread and examined as a film, are amenable to grinding with about an equal volume of NaCl which is then mulled in the usual way. Elastic materials are best examined by a completely different technique such as ATR. If this apparatus is not available, elastic solids can be ground, when cooled sufficiently, and examined as mulls (or pressed disks). Dry ice is the most convenient cooling agent, followed by liquid nitrogen; both are inert chemically.

Small, hard samples are not easily mulled without severe transference losses to the spectrometer cell. Grinding with NaCl is one means of 'expanding' small sample volumes, although it will not improve spectral intensity. Szonyi and Craske's approach is different and elegant.[185] A simple solution in a volatile solvent was added in small portions to a small drop of paraffin oil, each addition being allowed to evaporate before the next one. A fine sample precipitate was obtained in the oil, suitable for direct examination as a mull. Provided reference standards are treated the same way, this method could find application outside micro-work.

2. *Mulling*

The average published method, where a mull is described in detail at all, suggests adding one drop of paraffin oil to the ground sample, grinding further, and then adding further oil until the mixture has the consistency of a thin cream. One drop of oil is already *too* much. The average drop is approximately 50 μl and about five times too much. Better quantity control is achieved by using a scalpel to obtain a small oil scraping from the oil bottle stopper and smearing part of this scraping on the pestle which is then used to grind the sample further and thereby mix the two. Further oil can be added as required in the same manner until the whole solid has been moistened. Grinding is continued until a thixotropic product is obtained, a process which can be hastened by 'chasing' ground sample in the mortar with the same (wiped clean of excess oil) scalpel. The purpose of grinding under oil should be no more than a final finishing grinding and is really only a mixing operation. A correctly made mull will have a consistency more of a processed cheese, rather than cream normally described, with microlitre amounts of oil making all the difference.

Stanfield, Sheppard and Harrison[179] describe mull preparation in more detail than average, and their description of the end product as a thin ointment is as good as any. A mull is, after all, only a form of ointment. Their recommendations on grinding times err on the side of safety and should be adopted when in doubt of grinding efficiency.

3. *Transfer of mull to a cell*

Transference of the small quantity of mull in a mortar to rock salt disks can be tricky with a spatula or razor blade.[48,179] The by now familiar scalpel is preferred because of its sharp edge and sensible handle. Regrinding the cutting edge to fit the contours of the mortar is not normally necessary. A thin strip of mull, about 8×1 mm, is spread on each of the rock salt disks used as a cell, and the two disks placed together so that the two mull strips come together. Air inclusions are rare by this technique. At least three-quarters of the mull is recoverable from the mortar, and as the rock salt disks are gently squeezed together with *fingers only*, a film of mull is spread between them having an approximate rectangular area of 15×8 mm. This corresponds to the beam size at the sampling position of the average spectrometer and means that virtually all the sample is concentrated in the beam. The rock salt disks will retain their relative positions indefinitely and can be mounted in any convenient manner—suitable holders are invariably supplied as standard accessories with commercial spectrometers, and nearly all have quick-release fastening nuts. These nuts are tightened only as far as is necessary to retain the disks in position in the instrument AND NO MORE. Disks are fragile and crack and fracture easily. Cracked rock salt disks are rare with well-designed adaptable cell mounts (see Plate 4).

Common Faults

1. *State of grinding*. Grinding is common to both mulling and pressed disks and pays similar penalties for inadequate attention. Large sample charges cannot be ground efficiently, a practical limit in the average 35 mm mortar being about 5 mg. 3 mg is ample and yet 10–20 mg is not an uncommon sight, perhaps on the basis that twice the sample will give twice as good a spectrum, *ad infinitum*. This large charge of 10–20 mg will receive a token grinding of perhaps as many seconds, followed by another similar token grinding after the liquid addition. One minute's dry grinding of 3 mg is often not enough. Once liquid has been added, grinding efficiency is considerably reduced and appreciably greater time and effort must be expended to reach an acceptable particle size. The properties of 'wet or dry' emery paper are an obvious analogy. One of the reasons for this is the tendency of a liquid to flow away from a constraining force, and Cross in his book describes this effect well.[48]

Methods specifying the addition of liquid before grinding is complete are therefore inadequate; a homogenous suspension (cream-like) is reached long before particulation is complete. Some of the mull spectra reproduced by Cross show evidence of insufficient grinding. Other examples can be found in the *IR Atlas of Narcotics* compiled by Levi, Hubley and Hinge.[108] In this atlas nearly all the mull spectra have badly sloping base-lines and distorted band shapes at short wavelengths. Comparison with their spectra of narcotics in chloroform solution, or with other published spectra, makes this point graphically. It is of interest that these authors reject manual grinding in favour of a mechanical aid in an earlier paper in the same journal.[91]

2. *Mull quality*. The effort put into correctly grinding samples is all too often wasted by making a poor quality mull. A stiff mull can be thinned but the reverse is next to impossible with a limited amount of sample. Vast excesses of oil are all too frequently added by the tyro and in the absence of detailed instruction, the error is never corrected. Where published methods give any clue on transferring the mull to a cell, they invariably specify spreading it on one disk only,[48,179] and some manual dexterity is needed to avoid the inclusion of air bubbles. In an effort to overcome air bubbles, and reduce the skill needed, most workers take the line of least resistance and dilute the mull so that it flows more readily, and are caught in a vicious circle.

Another feature of a dilute mull is the large amount left in the mortar because of the limited quantity held between salt disks. Surface tension will be more effective with a free flowing oil, thereby holding the disks very close together and reducing the sample in the beam even further. If the sample is poorly ground, it is possible for large particles to act as 'spacers' while desirable fines are carried away by the oil and out of the beam, and thus accentuate spectral errors caused by large particles. Even a slight overtightening of the cell securing nuts will express more mull. Apart from aesthetic considerations, the risk of cross contamination in subsequent handling from this expressed material should be obvious. The nett result of a dilute mull from these interrelated factors is a poor spectrum with little more than major bands being observed and obtrusive dispersing medium bands.

At the other end of the scale, too stiff a mull will not spread evenly over the cell leading to excessive light scatter, and distortion of relative band intensities by the wedge effect.[101] Quantitative estimates are difficult. Major bands are frequently off-scale and swamp nearby fine structure. Hard samples have been known to scratch NaCl disks during attempts to spread stiff (dry) mulls, and particularly during subsequent dismantling.

The balance between too stiff and too dilute a mull is delicate. However, it is simpler to master than the alternative of using spacers between the disks to retain a respectable mull thickness for presentation to the sample beam. A correctly made mull will always give a thick film which can be reduced as needed to yield a spectrum of optimum intensity and detail. Some spectrometers are more sensitive than others and need a thinner mull to avoid swamping the spectrum with off-scale bands. It is always better to interpret 'thinner' in terms of film thickness in the cell rather than mull consistency and to avoid overemphasizing suspending medium bands, and a convenient balance between the two factors is best found by trial and error. As indicated on p. 41, over stiff mulls can, in the last resort, be thinned *in situ*.

Entrapped air bubbles are always a problem whatever quality mull is made, and become more difficult to eliminate as the mull is spread more thinly on the disk. Air is normally trapped because the second disk is not wetted by the suspending medium, which has a high viscosity and cannot flow over the second disk quickly enough. A very thin film on the second disk would be enough and, as indicated on p. 41, this is achieved in practice by placing a small drop of liquid on it (in the example cited, via the other disk). The same principle is applied in sandwiching the original mull. By spreading each disk with mull, the surfaces are wetted and no air bubbles between mull and disk are possible at this stage. What bubbles are formed occur in irregularities on the surfaces of the two strips as they are placed together. Since the thickness of the combined strips is up to 1 mm, and the width possibly only 1–3 mm in places, these small air bubbles are simply pushed to the edge of the mull as it is squeezed out. A small point, but one saving evocation of the deities, assuming, of course, that the operator has ever concerned himself with the elimination of air bubbles. Too dilute a mull will not build up into a deep enough strip, and a stiff or dry mull will not wet surfaces well enough to allow air bubbles to flow outwards.

Storage

Mulling is the technique of choice where ample sample is available because of its simplicity and speed. Storage of a prepared mull is therefore seldom necessary as it is quicker to remake a mull than laboriously recover, bottle and label the original. Another consideration is the ever present possibility of unique changes occurring in the sample. One is no better off if the re-run spectrum shows some differences from the original, comparisons only being valid for spectra run under comparable conditions. The original spectrum should be available for reference purposes in any event. Occasionally it does become necessary to store unique mulls, if only against the possibility of further chemical work being needed on the sample itself. Direct physical removal from NaCl disks can be a hazardous process for the disks. A metal straight edge is an obvious tool and is not recommended unless the operator has the necessary combination of steadiness of hand and luck. Each laboratory normally has to hand a number of ruined NaCl disks, which, if not already broken in two in a simple fracture, can be broken in this manner. The fracture is then ground flat, if not straight enough, and the straight edge formed used as a convenient scraper. The advantage of this approach is that scratching of polished NaCl is unlikely with the lubricant properties of the mull against a material of the same hardness. This assumes, of course, that undue pressure is not exerted. Too stiff a mull will tend to cement disks together, and is difficult to remove for the same reason, whilst too dilute a mull, or one containing a high proportion of large particles, will allow a good proportion of the suspending oil to creep away by capillary action leaving a stiff mull with similar clinging properties. Good mull preparation is therefore important if it must be recovered. Perhaps the safest scraper of all is a straight edge made from PTFE sheeting. Mull can be transferred to a small specimen tube by simply scraping the straight edge across the mouth of the tube, or in a two-stage process using a scalpel. Spatulas seldom have thin enough edges for the delicate and near quantitative transfer needed. The scalpel used in the mull preparation can be used to remove final traces from the preparing vessel.

Samples are not always soluble in the same solvents as the suspending oil, but where a convenient solvent, or mixture of solvents, is available, the mull can be dissolved off and stored as a solution. Evaporation of the solution will reconstitute an approximation to the original mull. The possibilities of crystal size change, solvation, polymorphism and residual solvents are endless, but as a technique it is worth consideration where a reference standard can be given the same treatment. Preparation of a mull by co-evaporation is not common as it defeats the essential simplicity of mulling, and can lead to spurious spectra as indicated. However, a micro preparation technique using only a sample solution has already been noted.[185]

POWDERS

The original solid sampling technique was for powders. It is little used today because of the inevitable severe energy reflections and scatter, and because of the simplicity of mulls and pressed disks. Provided the sample is fine enough, a recognizable spectrum can be obtained by the simple expedient of sprinkling it directly on a NaCl disk. Some allowance for radiation scatter can be made by attenuating the reference beam with a fogged disk or a polished one spread with fine NaCl powder, and for crystal orientation by rotating the disks.

It remains the only simple examination method for infusible solids in an uncontaminated state. No allowance need be made for matrix/sample interactions, or polymorphism induced by grinding if the sample is already finely divided in, say, a precipitation process. Considerable time can be saved in adopting this simple approach more often, provided the sample is divided finely enough. Devaney and Thompson[62] adopted this method to examine uncontaminated cured epoxy resins. Their grinding tool was a diamond-surfaced spatula, and their sample a manageable lump. This latter criterion is infrequent in analysis, but the application of two diamond-surfaced spatulas as a grinding tool for preparing powders is not so easily dismissed.

SUSPENSIONS

Although little known and rarely used, the suspension technique of Dolinsky does have application to the occasional intractable sample.[66,67] Suspensions are similar in their physical properties to mulls, whilst having many of the desirable properties of a solution, in particular, quantitative applications.

Sample is suspended in one of the usual solvents for infra-red by aluminium stearate, which appears to act both as a suspending and a solubilizing agent. The optimum concentration is regarded as 10 mg sample/ml in 1 per cent aluminium stearate in carbon disulphide or chloroform.

The required degree of dispersion was obtained in about 4 minutes using steel balls in a modified Wig-L-Bug amalgamator capsule having a capacity of 3 ml. No radiation scatter was observed below 1430 cm^{-1}, and adequate compensation could be achieved by balancing with a blank solution.

Dolinsky advocated the method for examining solids insoluble in non-polar solvents. Carbon disulphide and chloroform are not strictly non-polar from a chemical point of view and possess strong infra-red bands of their own in valuable portions of the spectrum. The main value of mentioning this technique at all is its potential wide application beyond the original intention. Attempts to make a 'normal' mull with hydrocarbon solvents do not appear to have been tried using aluminium stearate or a similar emulsifier. At a 1 per cent concentration, the emulsifier will make only a slight contribution to the overall spectrum and can be balanced out by filling a similar cell in the reference beam. Dispersion is best by mechanical shaking and should appeal to devotees of the non-manual approach.

LAMINATION

Published in the same year as the pressed-disk method, the lamination technique suffered the misfortune of being overshadowed by the more adaptable technique, and is little used. Lamination, or impregnation as it is also called, is essentially a special version of the pressed-disk method, and utilizes as matrix any transparent material capable of being formed (or obtained) in sheets. Sands and Turner in the original publication cite its use for plastic materials.[169] This provides the clue to its fall into disfavour. Crystalline materials are difficult to manipulate by this method, whereas pressed disks cope equally well with crystalline or plastic solids.

A further micro modification is given by Stewart[182] in which samples are compressed between two NaCl windows. It is described as being successful only for plastic solids. The sample is mounted between the two NaCl windows and squeezed by tightening the mounting screws until it is clear! The physical properties of the solid need to be well understood if the cracking of a large number of disks is to be avoided. The obvious rider to this approach is a sequential pressing—press a thin halide disk, place a thin and even layer of sample on it, cover with further halide powder or a disk, and repress. Relatively opaque disks are obtained with the properties outlined in the section on the nature of the sample (p. 18). Very intense spectra arise through the large amount of sample in the beam and provide an excellent way of studying minor bands; major bands are off-scale. When laminae are properly made, their spectra will show little base-line drift due to light scatter, and although the base-line may be no better than 10 per cent transmission, simple attenuation of the reference beam will bring the 'windows' and associated minor bands back to a usable part of the chart. Sample/matrix interactions are restrained to a minimum (at the two interfaces) and to avoid ion exchange the correct halide is obligatory. AgCl is quite a good matrix because of its own plastic properties, and is one of the materials mentioned by Sands and Turner. In the absence of pressed-disk apparatus (see Plate 5), adequate pressure for AgCl can be reached between the jaws of a bench vice, mounted vertically, or a large 'G' clamp, utilizing in either case aluminium sheets as the 'die'.

Lamination is always worth considering as an alternative to mulling where pressed-disk apparatus is not to hand. Plastic solids are not conveniently incorporated into paraffin oil by grinding. A thin film of solid is pressed onto the mortar or vessel and is not mixed by simple grinding, oil simply riding over it as a lubricating film. 'Chasing' with a scalpel is eventually efficacious but can be tedious. These plastic solids are more common than generally acknowledged and are ideal candidates for lamination. Amphetamine and methylamphetamine hydrochlorides are two such plastic materials, although they will mull without undue persuasion.

REFLECTION

Thin Films

Thin films on reflecting surfaces are capable of examination by infra-red. Energy losses are considerable on reflection and limit the application to specialist consideration in view of the more efficient ATR technique.

Applications have been discussed by Rappaport,[158] Robinson and Price,[167] and Giovanelli.[81] Cadman[23] in a more recent paper gave examples of the use of thin films in the examination of micro samples, mainly pesticides, collected from a gas chromatograph. The main difficulties arose from 'stray light', artifacts, polymorphism and sample non-uniformity. Surface reflection and internal reflections can be rendered negligible by placing an anti-reflective cover plate over the sample.[54]

Attenuated Total Reflection

First discussed in 1959 and published in 1961,[73] attenuated total reflection (ATR) is now a subject in its own right. Excellent commercial apparatus is available with adequate practical instrumental detail for all normal purposes (Plates 6 and 7).

Virtually all samples including aqueous mixtures are capable of examination with the apparatus, the only drawback of which is the initial high cost, most of them involving the use of the multiple reflection modification.

Pharmaceutical applications are given by Warren *et al.*[197] References 205 and 206 are recent reviews by Wilks, whose name has been associated with the technique since its inception.

Surface effects are artificially enhanced and care is therefore needed in interpreting spectra.

FILMS

The previous six solid sampling techniques all require some degree of manipulation and preparation for maximum efficiency and, between them, cover all methods in common use. The seventh and final technique, films, has been left to last because in its normal usage it is so crude as scarcely to merit description as a technique. Film technique is usually the description reserved for the examination of polymeric and like materials cast as films, and examined as such or in contact with an inert support. It is not the meaning intended here.

Liquids are easily presented as a film on a NaCl disk, although, in practice, this is rendered more stable as a capillary film between two disks. Semi-solids are similarly treated, mulls being a special subdivision. Solids are not so simply made into films directly, but a solution on evaporation will leave an even smear which is quite suitable for direct examination. Direct evaporation into the atmosphere at room temperature is most common, without any attempt being made to control the rate of evaporation. In spite of this crude approach, the film will dry in an amorphous or micro-crystalline form which will yield an adequate spectrum for routine monitoring. It is an old technique. Furchgott describes it for obtaining steroid spectra;[79] more recently, Mills applied it to the examination of 6–10 μg quantities of heroin using a specially grooved NaCl plate.[142]

The importance of this simple approach lies in its wide application to materials of forensic interest. Cross discusses it in almost as much space as his total description of pressed-disk preparation.[48]

Sufficient solid is dissolved in a simple organic solvent to deposit enough for a recognizable spectrum. Chloroform is generally the preferred solvent because it dissolves amine and alkaloid hydrochlorides to an appreciable extent. The alkaloid content of tablets and capsules is rarely very high and requires careful chemical analysis to conserve sufficient for duplicates or further confirmation. A simple chloroform extract deposited on a rock salt plate may often give an immediately recognizable spectrum, even in mixtures, and save considerable chemical work. Illustrations are given on pp. 103, 115, and 153.

Used in this way, this method provides one of the few occasions where infra-red spectroscopy can be applied as the first action method as a diagnostic tool. The sensitivity can approach that of a pressed disk without elaborate preparation and, in addition, the smear is uncontaminated and may be washed off with the same or other solvent for chemical study. With a good retrieval system, positive answers are obtainable in minutes, and as the answers are for all intents and purposes nondestructive, quantitative assays by this or some other method can be completed with the confidence that the correct material is being analysed.

At the other end of the scale, there is little point in isolating trace amounts of a substance and then subjecting it to the hazards of mulling or pressed disks if a simple film will suffice. Apart from gas liquid chromatography, chromatographic fractions are eluted as a solution of some sort, and are in a suitable form for direct evaporation onto any simple cell window, assuming, of course, that the solvent will not dissolve the cell window.

Overall sensitivity is extremely dependent on the nature of the sample and deposition conditions. Thin, even films of an amorphous solid dissipate little incident radiation by scatter and relatively strong spectra are obtained, whereas discrete, and large, crystals scatter an appreciable amount of energy and require some compensation. The best compensator is a fogged NaCl disk of the type used for mulls since this has thc desirable effect of 'straightening' the base-line so that the end product is much more amenable to scale expansion.

Some materials with a low melting point are suitable for a modified film technique. Simple melting will spread a thin film over any available surface like a NaCl disk, although it is sensitive to thermal shock and will need some device capable of a gradual heating and subsequent cooling. Most laboratories possess a Köfler block or its equivalent, an ideal piece of apparatus; the risk of oxidation or decomposition is high and in practice the substance is heated between two NaCl disks to exclude air as far as possible. Polymerization is another cause of anomalous spectra and may be noted as a simple change in certain bands; however, if a single crystal in film form is produced, polymerization may cause some really weird effects. Infra-red radiation is polarized in most spectrometers, and the relative intensities of some bands will be altered by rotating the sample in the beam . . . It may be pure coincidence that the band chosen to line-up the overall intensity is one subject to fluctuations due to polarized radiation . . . The safest answer to this problem, because of its unpredictability, is to record spectra with the cell rotated through 90°, or to forget melted films as a technique altogether. Unidirectional orientation of all crystals by the evaporation version is highly unlikely, unless an amorphous film recrystallizes, and is usually ignored as a problem.

A simple drop of solution allowed to evaporate naturally, whilst standard practice, is open to five other objections:

1. Rapid evaporation leads to crystal formation rather than the preferred amorphous film.
2. The deposit is uneven, with a major concentration along the edges of the drop. Rotation of the cell for maximum absorption is not an entirely satisfactory expedient to overcome this.
3. Rapid evaporation causes moisture condensation from the atmosphere. Samples may be altered in chemical form by the water or, in extreme cases, the cell material will pick up enough water to be seriously damaged by fogging. The immediate effect will be obtrusive water bands in the spectrum and increased radiation scatter particularly at short wavelengths.
4. Delicate sequential application of drops and subsequent control of evaporation are tedious and time consuming. Strong solutions need only be applied in one drop; dilute samples require many. When the time taken to prepare a mull or pressed disk is taken into account, this objection would appear to be of marginal importance. Considering the time and effort needed in cleaning apparatus used in mulling or pressed disks, this fourth objection shrinks to its true perspective.
5. There are always awkward samples which prove difficult to remove subsequently from the disk. Soaking the cell in a shallow layer of solvent in a beaker is required to loosen the layer for subsequent removal by fresh solvent. Fortunately, nearly all samples are removed by slowly dripping solvent over the smear, and allowing it to drop off the bottom edge of the disk into a suitable receptacle.

A very neat method of overcoming all of these objections was proposed by Huo-Ping and Edwards.[92] The cell window is placed under a hair dryer and solution applied to it using a pad of purified cotton wool held in a capillary pipette. The pipettes are manufactured from ordinary glass tubing and are disposable. There is a double advantage in this approach, in that solids in suspension are filtered automatically and held in the pad. In these cases the pipette contains two successive pads, and after filling, the wad near the tip is broken off at a mark previously made with a file. My own preference is to use a filter beaker with, usually, a number 4 sinter to reduce the filtration rate to the evaporation rate on the disk. In either case clear solution can then be applied as before. Tablets and capsules can therefore be examined very simply and rapidly without recourse to any complex apparatus, all that is needed being a small ignition tube, the disposable pipette applicator or a filter beaker, and a single NaCl disk. With very volatile solvents such as chloroform, even the hair dryer can be omitted.

As a general rule, spectral intensity is low with the film technique and some degree of scale expansion is necessary with the smaller samples derived from chromatographic separations. The characteristic feature of most film spectra is a very flat base-line with a few peaks protruding from it, very much as though the bottom portion of the spectrum on the chart paper had been sliced off with a guillotine. This very crudeness is a distinct advantage for simple indentifications, the half dozen or so strongest bands are the only ones seen and are therefore readily picked out without obscuration or intensity modification by minor bands. Retrieval schemes rely on a positive identification of the strongest bands, and fine detail is of no consequence at this stage. Unique answers are not impossible from spectra which would be quite unacceptable to a theoretical spectroscopist.

Caution must be exercised in applying this method. Polymorphic differences are the rule rather than the exception, and controls must be run for every material encountered for the first time. Examination in this way as an evaporated film is the simplest way of investigating polymorphism. Bulk supplies of authentic samples are usually recrystallized from aqueous solutions, whilst films are deposited from a variety of organic solvents. Spectra bearing no immediate resemblance to those in a mull or pressed disk, or published in the literature, must always be anticipated.

2

SOME OTHER BASIC TECHNIQUES

Think of ease, but work on.

GEORGE HERBERT

RECORDING A SPECTRUM

Concentration

After placing the mull or pressed disk in the spectrometer, it is necessary to check one or two points in order to obtain an optimum spectrum. Manual scanning will locate the strongest bands (which should not invariably be due to the suspending medium in a correctly made mull) and the wavelength setting should be left at one of these. The ideal absorption for the strongest band is between 80 and 90 per cent and is not always obtained first time. Some unevenness is to be expected in mull distribution across the cell and a control over band absorption can be achieved by rotating the cell in its mounting, provided the mull continues to fill the beam. In cases where absorption is too much, it can be reduced to some extent by *slight* tightening of the cell fastening nuts. There are obvious limits to this approach, the best answer being experience of the quantities most suitable for the spectrometer/cell mount combination. Remaking the mull is not always feasible and in extreme cases of strong absorption, the two salt disks can be slid apart, one wiped free of mull, and a small portion of oil added to the other. Careful closing again from one edge (clam-shell-like) gives an even chance of completing the operation without entrapped air bubbles. Gentle rotation of the disks relative to one another will remake a thinner mull, which, in time, will become even over the whole disk area. The concentration in a pressed disk can only be altered by remaking the disk at a different concentration or thickness.

Once sample thickness is acceptable, attention can be given to background absorption.

Compensation

Spectrometer design varies so much that it is not possible to give more than general rules. Where sample preparation can be rendered reproducible, the background absorption could be taken as a sample characteristic and no correction applied. Alternatively, the worker may be unaware that such a correction is applicable! Although good sample preparation will give high transmission, there is always a small amount of energy scatter which lowers the overall band intensities, and this is particularly true when there is a bad mismatch of refractive indices between sample and suspending medium. Some of the cheaper instruments do not provide more than a crude sample beam trimmer, which, in a number of cases, is invariably set at its maximum the whole time.

Several alternatives present themselves to achieve the necessary reference beam attenuation.

1. *Reference beam attenuator*. Use the instrument's reference beam attenuator where this is fitted, or, alternatively, purchase one and use this. Plate 9 illustrates a simple but effective attenuator.
2. *Metal gauze*. Crude adjustments are possible with a selected range of gauzes. Rotation of a gauze in the beam, or in an axis perpendicular to the beam, gives an even cruder fine control. This is not a very elegant approach but quite efficient and cheap.

These two alternatives make an allowance for the overall background alone. The next three compensate for radiation scatter which can be appreciable at short wavelengths.

3. *Fogged NaCl or KBr disks*. These make some allowance for the absorption characteristics of the disks used with the

sample as well as correcting for light scatter. Often it is suggested that a 'ruined' disk be relegated for this purpose. It is better to take a brand new unpolished blank, to polish one side reasonably well, and to polish the other carefully until the required degree of compensation is reached. Deep scratches made by the manufacturer's rough grinding are not removed and allow a small range of fine adjustment by rotating the disk. Once made for an average sample, these disks will serve as a permanent reference attenuator, NaCl for mulls and KBr for pressed KBr pellets.

4. *NaCl mull.* This is the best compensator available for mulls. Overcompensation is the rule rather than the exception and very careful adjustment is necessary to avoid sharp negative spikes on the spectrum.

   The main advantage of a NaCl mull is the degree of compensation possible for the suspending medium. An alternative is a simple film of medium between two NaCl disks although this method tends to form too thin a film for complete compensation. Because both approaches require some degree of expertise for optimum efficiency, the method is not often encountered.

5. *Pressed pellet.* Method 3, using a KBr disk, is normally more than adequate for the average sample. However, some pellets scatter more energy than others, and the simple fogged reference disk is unable to cope with the excessive energy scatter compensations needed at short wavelengths. Also a KCl disk is virtually unobtainable for compensation and the only answer is to press a reference pellet. Variations in the applied pressure and the quantities taken for pressing give a considerable measure of control over the light scattering properties of the reference pellet. As noted elsewhere (p. 15) the Mini-Press is an ideal tool for this delicate work.

   Apart from the precise compensation possible with a custom-made reference pellet, there are the added advantages of impurity compensation in the matrix and compensation for other unwanted components in a sample mixture.

Whichever compensation method is chosen, adjustment is made at a point on the spectrum seen to have a minimum absorbance. Usually this is around 2000 cm^{-1}. No amount of reference beam attenuation is an acceptable substitute for good sample preparation, but, as has been noted in the previous chapter, there are always occasions where sample and matrix are incompatible and an 'opaque' pellet or mull arises through no fault of the worker. With a relatively low, fixed, refractive index matrix, light scatter can be appreciable and is the most common reason for the phenomenon. Bringing the point of maximum transmission back to 95 to 100 per cent can have a dramatic effect on the overall band intensities, giving the appearance of an amplification. Although this is a practice to be frowned on from a theoretical point of view, with all the variables involved in grinding, it is a convenient method of enhancing a basic pattern for identification purposes. It also ensures that virtually the whole chart paper is used and not just x per cent, a factor which often loses some minor bands into the background, thereby losing potentially useful diagnostic information.

Similarly, if methods 3 to 5 above are used, the apparent baseline drift due to light scatter, particularly at short wavelengths, can be corrected or eliminated entirely. This becomes very important if the information/spectrum retrieval method relies on band intensities or retrieval is based on bands showing an apparent high absorption, that is, where bands are indexed according to their closeness to the 100 per cent absorbance line, rather than characteristic/functional group presence.

BEAM FILLING

Unless the sample completely fills the beam, a weak and possibly distorted spectrum will occur.

Pellets

Provided pressed pellets are mounted in an opaque mount, beam filling problems are uncommon if the pellets are uniform in concentration, as they ought to be. A simple mask is normally all that is required to avoid phenomena associated with an incompletely filled beam.

Mulls

Mulls require more care in preparation than pellets. If the sample is in short supply, there may be no alternative to a mull incompletely spread over the cell window and making the best of the situation by rotating the cell, or its mounting, until maximum band intensity is obtained. This is one way of overcoming air bubble problems and is the normal remedy. An excessive number of air bubbles will not only produce a weak spectrum by reducing the amount of sample in the beam, but will also increase light scatter and alter the apparent base-line. Ordinary small samples require a mask to restrict beam size to the area covered by the sample; the reference beam is similarly masked.

The converse situation is more serious. Too stiff a mull will not spread fully and yet will give strong absorption bands. The bands can often be brought back on-scale by rotating the cell and a spectrum recorded. Relative band intensities are frequently incorrect in this technique, with truncation of major bands—the appearance is as though the whole spectrum had been compressed into a narrow band of the chart. Mild cases often pass unnoticed and only become important where a number of strong bands of approximately equal intensity could be selected for data retrieval.

The factors of mull quality and beam filling are interrelated. A good quality mull needs no special treatment whereas a poor example needs some dexterity in order to utilize the whole chart. Whether too thick or too thin, the results are similar in reducing the diagnostic value of minor bands; too thick and minor bands are over emphasized in relative intensity, too thin and minor bands are lost into the background.

POLARIZATION

The effects of polarized infra-red radiation on the spectrum are not always obvious with ordinary mulls and pressed disks, and are normally swamped by other effects. Solid films frequently suffer from polarization, and with these samples it is safest to assume this to be the rule rather than the exception.

Provided the sample is well dispersed in the suspending medium, polarization phenomena are unlikely, but should not be dismissed entirely, particularly in a mull, if the sample is present in large crystals, through inadequate particulation, or is a polar molecule. Spectral effects due to large particles are discussed in Chapter 3, and most of these are sufficient to invalidate the spectrum anyway without reference to polarization. With a well-ground and polar molecule, there is always a possibility of micelle formation which may polarize incident radiation.

Rotation of the cell has been mentioned in various parts of this book as one method of altering spectral intensity in a non-homogenous sample. Rotation through 90° is a worthwhile precaution to confirm that polarization phenomena are not obtrusive, at least for each compound encountered for the first time. This precaution becomes mandatory, even with pressed disks, where quantitative work is envisaged.

Micelle formation in a liquid phase appears not to be a problem, and quantitative measurements on a polarizing substance should be carried out in solution where there are no suitable alternative bands immune to intensity fluctuation in the mull or pressed disk when the cell is rotated.

MICRO-WORK

Micro-work is inevitable at some stage in analytical chemistry.

Once the apparatus and techniques have been acquired, the time taken need be little more than that taken with a bucket of sample. Commonsense is essential to achieve good results, but beyond this little more than scrupulous attention to cleanliness and detail is needed. The techniques have become so familiar and wide-spread that there is some strength in the argument that semi-micro, if not micro, techniques should become routine anyway for the bulk of examinations. In the last few years improvements in instrument sensitivity, speed of response and optics have all played their part and few spectrometers are now offered without a standard range of micro accessories and scale expansion.

A small amount of sample in the beam necessarily means a small detector signal and two alternative approaches are adopted, often together, to improve the signal.

Scale Expansion

Where scale expansion is electronically derived within the instrument, the nett gain can be very slight. Unless the electronics are in good condition the base-line noise will be amplified disproportionally, a factor that has caused some embarrassment in the past, and occurs mainly where the electronics have been specifically designed for a particular machine and have, understandably, no more sensitivity than needed for the original purpose in hand. Amplification of a very small signal from such an amplifier working near its limit cannot do much more than appear to amplify the noise, because the recorder may have a time constant or damping circuit capable of averaging out most of the noise at its normal input, and thereby give a false impression of stability. A tendency for the recorder pen to move stepwise or jerkily, rather than smoothly, with low signals at normal settings, is indicative of the possibility of a large apparent increase in base-line noise on scale expansion, losing some fine detail into the background. Most modern spectrometers have good electronics and will stand up to a twenty-fold amplification in some cases.

A slightly different approach is to use a recorder with a larger chart size or a variable voltage input. A variable input recorder without an effective damping smoothing circuit will suffer from the same objections as electronic amplification within the spectrometer, particularly when its time constant is significantly smaller than the original recorder. As a technique, it is becoming popular in visible and ultra-violet spectrometry to 'stretch' the capability of some of the simpler spectrometers and will undoubtably prove efficacious with some infra-red spectrometers. The main difficulty with any external recorder is an accurate correlation with the wavelength scan. A change in chart size is rare since most recorders are limited to a scan of between 6 and 11 inches. The Hilger instruments are an exception and produce their range of charts and scale expansion by a mechanical rather than electronic device, effectively a variable span recorder.

Beam Condensing

In an attempt to increase the effective concentration of sample in the dispersing medium, the amount of dispersing medium is decreased. The nett result may be no better than that from the originally dilute specimen if the pellet/mull area remains constant. If the sample is presented in a small area to the incident radiation, a characteristic or 'clipped' spectrum can be seen cramped into the high transmission portion of the scale if a significant proportion of the radiation bypasses the pellet altogether. Blocking off all but the area covered by the sample, as in most micro die holders, gives a similar cramped spectrum, but in this case it is cramped into the lower transmission end of the scale. Attenuation of the reference beam will bring the spectrum back to its normal place on the scale, but there are limits to this approach. The reduction in energy passing to the detector is tolerated by some instruments better than others and attenuations of the order of 95 per cent are possible, although a limit of about 90 per cent is agreed in general practice. The function of a beam condenser is to concentrate the sample beam into the area covered by the sample and thereby avoid the necessity for large reference beam attenuations, and as the main beam is effectively 'seeing' a higher concentration, the spectrum is amplified.

Plate 10 shows a typical beam condenser based on refracting optics.

What is not generally appreciated is that an average beam condenser will only transmit about 50 per cent of the incident radiation and a low rate of scan is advisable to overcome an inevitably sluggish recorder response when the pellet already has a low transmission from faulty manufacture or sample incompatibility. During this long scan an abnormally high energy concentration is directed onto the sample and heating effects can be quite considerable, causing fundamental changes in the nature of the sample.

Special precautions may be needed to prevent evaporation of volatile solvents in solution work. One device is to place a separate cell with a thin film of the solvent in front of the micro-cell to absorb the bulk of the heat energy. Whilst effective, it will reduce the wanted radiation reaching the detector even further and make even more call on the operator's skill in compensating the reference beam. This is rarely used except in the most exacting or quantitative work, but the similar use of a heat filter is worth consideration in cases where thermally labile compounds in pellets are being examined for extended periods. An AgCl disk exposed to H_2S gas is a simple heat filter.

Other aspects of pellet heating are considered later (p. 72f).

There are two elegant methods in micro-sampling which deserve special mention as neither of them requires a beam condenser. Sterling[180] utilized a standard half-inch die press with normal fitments. A blank KBr disk produced by the press has a small hole drilled through its centre by a specially guided drill. Details of the jig are given. Prepared sample is placed in the resulting cavity and pressed in the normal manner into the blank disk, and being surrounded by this disk it is easily handled. Because the 'holder' is optically clear it will not attenuate the sample beam to any appreciable extent and the small amount the sample beam is attenuated is compensated for by placing a similarly prepared but blank disk in the reference beam. A weak spectrum typical of samples incompletely filling the beam is obtained in spectrometers of older vintage with broad focusing, whilst modern instruments with good focusing will yield an amplified spectrum. In either case the spectrum is more amenable to subsequent scale expansion because there is no appreciable beam energy loss involved.

Chen and Gould's approach is fundamentally different and uses AgCl as a working medium, employing its ductile properties to make an integral beam condenser in the pressing process.[33]

MICRO-CELLS

Chromatographic fractions are normally encountered as a solution because this gives some tangible volume to an already diminutive sample. Occasionally the product is liquid anyway. Alternatively, the interesting substance is not conveniently prepared as a solid, either as a salt or a derivative, and the time taken and potential losses in quantitative work preclude a simple application of solid-sampling methods. Micro liquid cells are an essential part of the spectroscopist's arsenal.

Commercial items are relatively cheap (very cheap if labour costs are taken into consideration), but technical journals contain a large number of very practical suggestions, some of which have become commercial ventures . . . and so the cycle continues.

One of the simplest ideas of all was given by Black.[16] A KBr disk was pressed in the normal manner with a small piece of platinum (or other) wire set in it. Withdrawal of the wire left a simple micro-cell which could be mounted in familiar accessories. A whole range of capacity will be available by simply selecting the right wire gauge.

Small blocks of rock salt can be bored if treated gently, and the cells of Glassner's design[82] are now a commercial item.[35] The more enterprising brethren may wish to manufacture cells for themselves to fit some specialist gadget, and will find the practical details by Price[156] on the art of boring, and by Nicholls[149] on cleaving small plates of some assistance.

Polishing cells is no problem to a person familiar with the manufacture of astronomical or optical mirrors where the wavelength of light is an order of magnitude shorter. The jig suggested by Feairheller and Du Four[76] is simplicity itself and quite sufficient, and, as the authors indicate, can be adapted to deal with windows that are not circular. The important point about this jig is that it depends on an even pressure applied by an inert object, and does not need a deftness of touch which most simiae can never achieve. Anything more elaborate is best left to an expert or a separate research project.

Price's method of mounting[156] is sufficiently novel to merit special mention. Few spectroscopists understand the inner optics of their charges and fewer still have the temerity to 'gut' them and place the micro-cell near the slit as Price does; it is a neat way of avoiding a beam condenser if you can steel yourself to try it.

Capillary tubing is an obvious source for infra-red micro-cells and has been extensively tapped. Polythene is manufactured in a wide range of tube sizes and has the bonus of being ready for use without tedious preparation. It is also available to a relatively tight specification, making it suitable for direct use as a reference cell with very little imbalance, whilst in the far infra-red, its spectrum can be ignored for most practical purposes and need not be balanced. Silver chloride is a very ductile material as well as being infra-red transparent. Blout[19] extruded it in lengths of 1 to 1·5 mm with a bore of 0·075 mm as he needed them. Cells based on this principle can be purchased.[72]

QUANTITATIVE ANALYSIS

Control of the parameters in solid sampling is relatively difficult and an assay by ATR or in solution is to be preferred. Solution techniques are well described in most textbooks, and ample experimental detail is invariably supplied with commercial cells, sufficient, in fact, to guarantee adequate success for the average person without a lot of experience. Quantitative ATR problems are discussed elsewhere (reference 206). Comments on quantitative aspects will therefore be restricted to solid sampling except to mention the short note by Clark.[37]

This was intended as a simple method for studying liquid samples directly. It is a simple and straightforward method with the advantage that solid-sampling apparatus (a KBr press) is used; with careful attention to detail in their manufacture, the cells could be applied to quantitative work without recourse to expensive commercial equivalents. If the cells are treated as expendable, cell cleaning and the possibility of cross contamination are avoided.

Only three approaches are selected out of the many possible quantitative methods.

Direct Assay

For a known single and pure substance, the simplicity of a direct measurement on one (or more) bands has much to commend it. Provided the band chosen is not subject to interference, quite complex mixtures can be treated in the same way; in either event, calculations are based on a known extinction coefficient (more applicable to solutions), or by reference to a standard curve. It is not impossible to use a mull quantitatively provided the working parameters are rigidly standardized, in particular, the thickness of the mull sandwiched between the cell windows. Apart from the obvious problems of non-homogenous distribution over the whole beam area, the nature of the sample can have an important bearing on the answer. Two crystal effects were noted a long time ago, and Barnes and his co-workers discussed in some detail the effects of crystal orientation and pleochromism on the quantitative aspects of mulls, in this instance with reference to penicillins.[11]

Control of the conditions with pressed disks is relatively easier but the reader must make himself aware of the influence of refractive index mismatches and particle size, and is referred to the review by Duyckaerts.[69]

Internal Standardization

It is frequently found more convenient to complete an assay by reference to an internal standard in an attempt to eliminate the influence of preparation variables. The major unknown is either the matrix thickness, or apparent amount in the beam, and this can be assessed from the internal standard. Lead thiocyanate is a favourite one, and may be employed with equal ease for a mull or pressed disk. Bradley and Potts,[20] and Resnik[161] are two references selected at random to illustrate the two techniques. Resnik gives more detail on its application to the assay of paper chromatographic fractions.

Differential Analysis

Although of more application in qualitative analysis, differential analysis is a powerful technique for assays in situations appearing hopeless at first sight. The only requirement is some knowledge of the source(s) of interference, and, in this sense, it is really only a refinement of the usual reference beam balancing for background absorption.

Some dexterity is needed to reach a precise balancing out of the unwanted absorption, but once mastered, it will give every satisfaction because of its universal application. Robinson[166] specified a limit of about 40 per cent transmission before applying such a correction, but, only a few years later, Washburn[198] recommended a procedure applicable down to 10 per cent transmission (for liquids in his examples). Modern spectrometers are sufficiently sensitive to return a reliable assay under these conditions of high compensation. The assay of tablets is greatly simplified by differential analysis since it can be carried out on the tablet without extraction of the component of interest. The major excipient in a high proportion of commercial tablets is lactose, which has a devastating effect on a spectrum with its strong, broad bands. A relatively concentrated lactose mull or pressed disk can be made and held in reserve for these cases, and by gradually increasing the amount in the reference beam balance out the lactose spectrum. It is normal to make a wedge-shaped disk for this purpose. Other excipients are generally present in small amounts or do not have such a 'swamping' spectrum. The determination can then proceed on the band(s) of choice in a normal manner. A natural limit to this technique for tablets is set by the concentration of substance in the tablet, which must be sufficient to yield a measurable spectrum, but, nevertheless, the technique can save a considerable amount of time with a high proportion of tablets and capsules if a high standard of accuracy is not required. Checking identity and level of contents is an obvious application, and if an authentic standard tablet is available, this could be used in the reference beam (prepared in precisely the same manner as the original) and a 'null' spectrum recorded as an extra confirmation of the assay. If it is not practical to duplicate the original, the wedge technique mentioned above could be used. Manufacturing tolerances can have an important bearing on the method if this is adopted, and must not be forgotten; the *British Pharmacopoeia* requires twenty tablets to be taken to allow for this factor.

The principle of differential analysis has progressed from the obvious applications above. It is not always possible to obtain a pure reference standard for compensation purposes, and often one of the impurities encountered in it is the substance of interest. A substitute standard was advocated by Washburn and Mahoney[199] and they cited four pharmaceutical applications with a precision of better than 1 per cent (0·1 per cent in one case). The same principle was adopted by Manno[120] who obtained a similar level of accuracy.

The important aspect of Washburn's and Manno's approach is that a relatively simple spectrum is seen, with the peaks in this standing out strongly derived from substituents characteristic of the test substance not common with the reference molecule. It is also applicable to the estimation of low concentrations, a real-life situation in forensic work. Both authors discuss aspects relevant to a choice of the substitute standard, and provide important clues for applications in other fields. The limit is set by the ingenuity of the worker.

Some of the mundane quantitative applications to pharmaceuticals are given at regular intervals in the literature, for instance in the *Journal of Pharmaceutical Sciences*. A typical one, in the form of a review, was printed in 1960 under the authorship of Hayden and Sammul, in which some barbiturates were considered in the examples.[87] Occasional pharmaceutically inclined papers appear in *Analytical Chemistry*, one of particular relevance being by Browning in 1955[21] concerning the determination of scopolamine and atropine in tablets.

However, before infra-red spectrometry is adopted as a quantitative technique, some serious consideration is needed on the question of its necessity in the first place in forensic analysis. The Drugs Acts in Great Britain specify that possession of 'any quantity' of a restricted material is an absolute offence. Therefore, at face value, it is not necessary to determine the quantity of restricted material present, and if not restricted, quantitative assessment is irrelevant. However, the quantitative aspects can be an extra very helpful analytical confirmation of other data, and provide the courts with a simple idea of the magnitude of the offence, although it must not be forgotten that the important criterion is the original gross weight, or number of items involved—a tablet is always intended to be a single, or even multiple, dose regardless of how minute its contents might be. Similarly, toxicological work may not appear to require accurate assays, but this evidence can be vital when related to the medical evidence.

Quantitative assays are frequently more accessible by a totally different method, perhaps one which will provide extra confirmation of identity. In this connection, solutions in water are simpler to handle and visible or ultra-violet spectrometry is more familiar and less subject to sampling errors during the preparation, and would appear to fill the bill admirably. Another familiar assay, which in this case is scorned too frequently, is a simple weighing—after an extraction, the extra time taken for an evaporation to dryness and weighing (as a salt for a free base!) is very little, and does have the advantage of being an absolute method. Modern balances, even on 1 mg residue, will weigh to an accuracy better than most requirements since 0·05 mg, or one-half division, is 5 per cent and not much worse than the claimed accuracy of an average spectrometer; most balances can do better than this without effort although they may not be describable as micro-balances.

These remarks do not decry the use of infra-red spectrometry for quantitative work—in quality control or allied fields it is invaluable—but the analyst requires a clear mind when considering the need for it, and must survey all aspects of the case before reaching a decision. Court evidence is required to be accurate with a greater degree of confidence than is generally necessary, and the return on the effort involved is often difficult to justify if another independent assay will suffice. Two court decisions have clarified the situation and emphasized the importance of absolute methods. In the first of these,[160] small quantities of a substance were detected with a microscope and subsequently found to contain heroin; the defendent was convicted of unlawful possession. An appeal was upheld on the basis 'That particles of a substance invisible to the naked eye and incapable of being measured, poured, sold or used are insufficient to found a charge of unlawful possession'. In a subsequent case, the appellant cited the earlier precedent, after conviction for possession, when minute quantities of cannabis had been found in his clothing. The case was dismissed on the basis 'That scrapings of cannabis from pockets, which, whilst minute, can still be weighed in milligrams, are sufficient to found a charge of unlawful possession'.[159] Cannabis is admittedly not a drug whose identity and assay are based on infra-red spectrometry, but the principle of a weighable amount being present has been established, and it is possible to refuse some assays, other than in general terms, with a clear conscience, provided, of course, one is not dealing with a reasonable request.

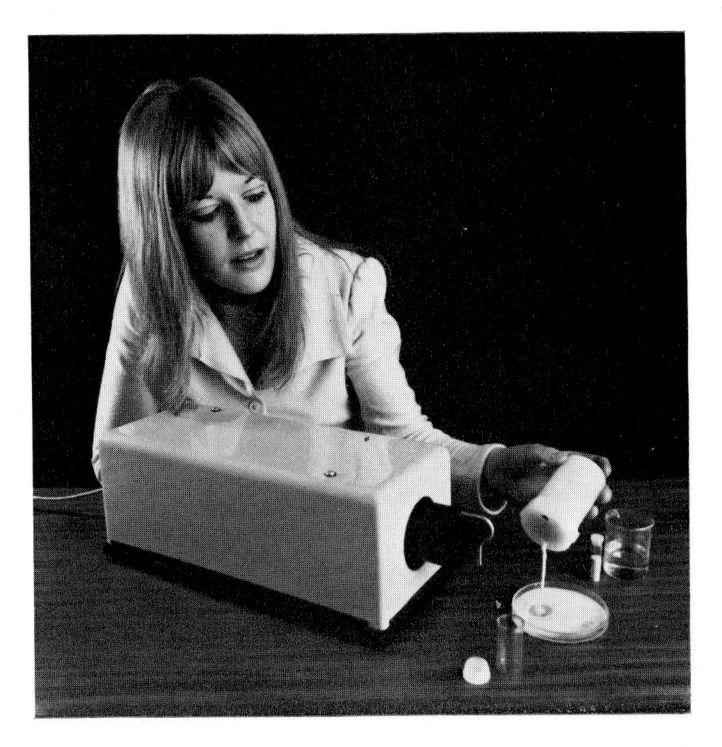

PLATE 1. A McCrone micronizing mill. It is robust, capable of handling relatively large quantities of sample, particularly hard materials. Contamination by plastics in preparing large batches of matrix should not be critical. (See p. 11.)

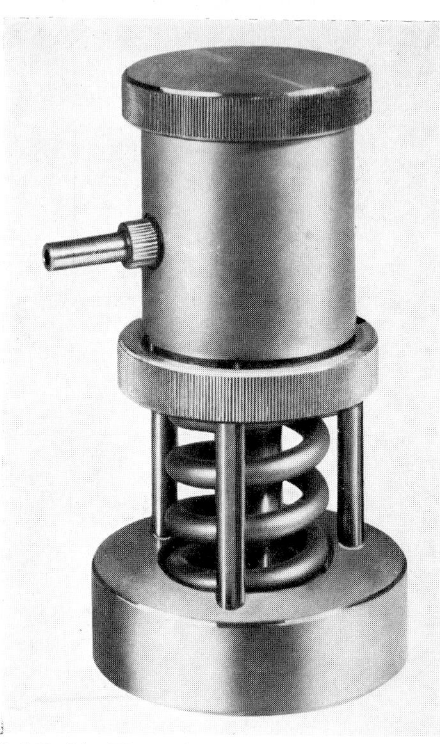

PLATE 2. A Perkin–Elmer ultra micro KBr die which produces pellets 0·5 and 1·5 mm in diameter. It has an automatic pressure limiter to prevent damage to the die. (See p. 15.)

PLATE 3. A micro KBr disk holder and prepared pellet (1·5 mm). (See p. 20.)

PLATE 4. Exploded view of a Perkin–Elmer demountable cell. The cell is suitable for viscous and non-volatile liquids and mulls, and is provided with a range of spacers for different cell window materials and thicknesses. (See p. 28.)

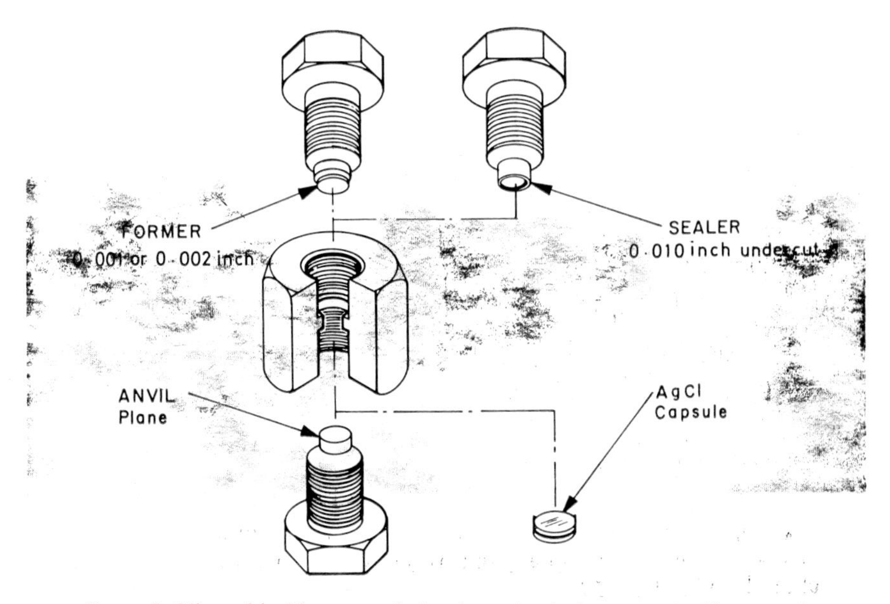

PLATE 5. Silver chloride encapsulation (examination) apparatus. (See p. 34.)

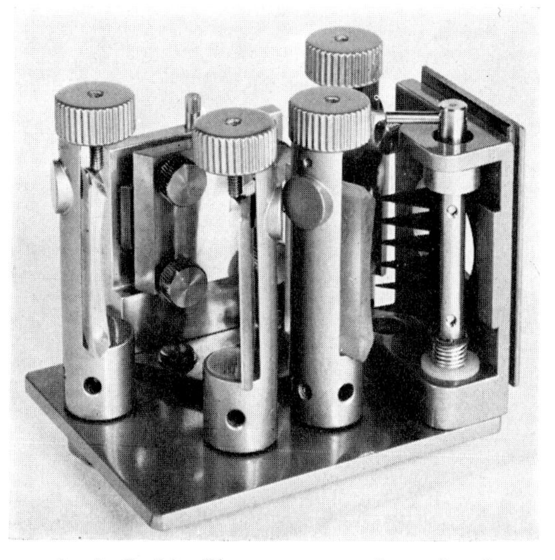

PLATE 6. A Perkin–Elmer attenuated total reflectance accessory. It can be supplied for three, nine, or twenty-five reflections. There is also a micro-ATR version of it. (See p. 35.)

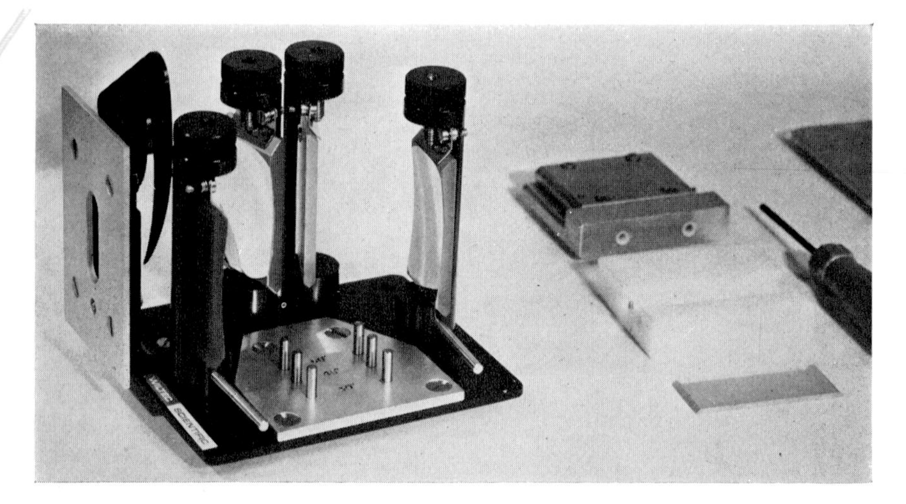

PLATE 7. A Wilks attenuated total reflectance accessory. The sample holder may be mounted in one of three positions (30°, 45° and 60°) to change the angle of incidence. (See p. 35.)

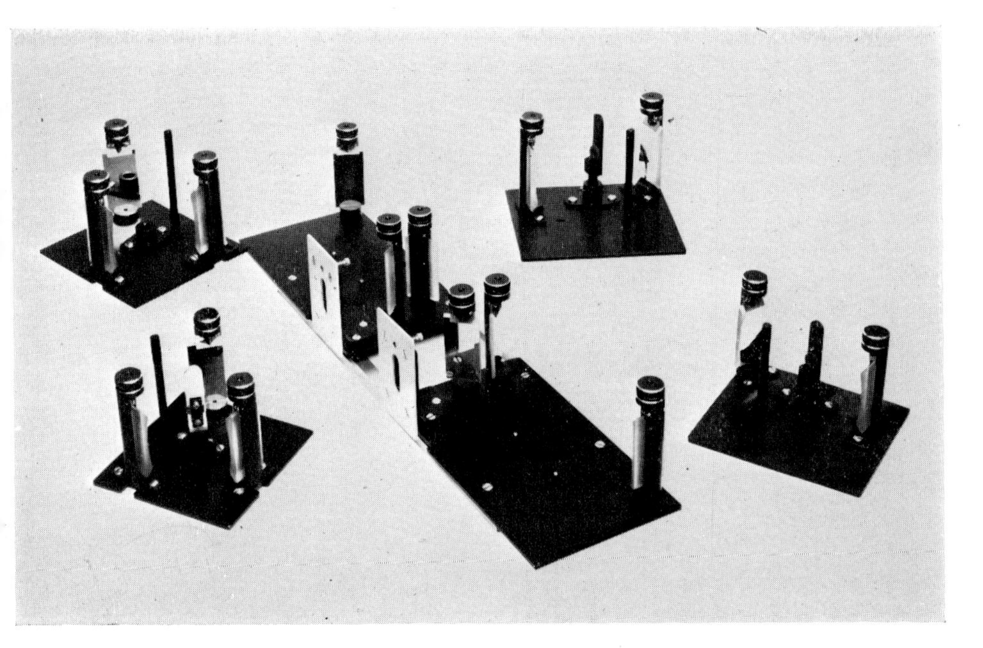

PLATE 8. Wilks universal micro-sampling system. The basic unit is a 4-to-1 reflecting beam condenser in a modular construction for micro-ATR, micro-transmission and micro-reflectance work. The complete double-beam unit (model 45DB) is shown here. (See p. 35.)

PLATE 9. A simple and effective reference beam attentuator manufactured by Perkin–Elmer. (See p. 42.)

PLATE 10. A Perkin–Elmer refracting beam condenser. The KBr lenses yield a fourfold beam area reduction. (See p. 48.)

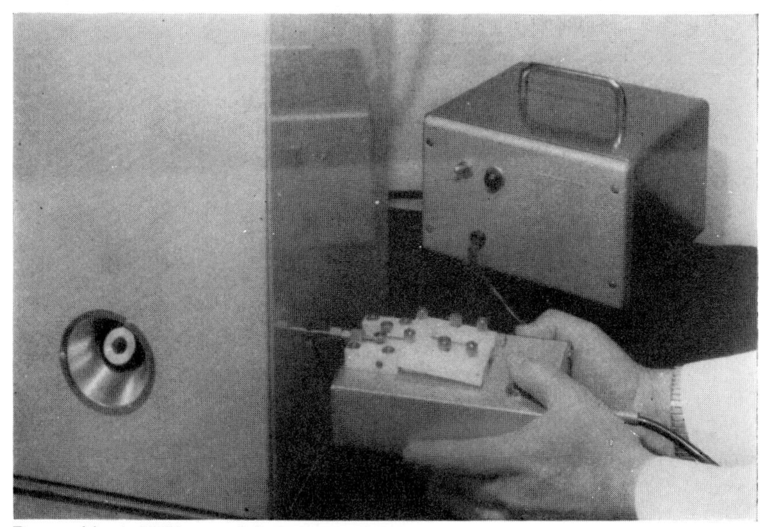

Plate 11. A Wilks model 15 for collecting gas chromatographic fractions by Pettier cooling (shown in operation) (See p. 76.)

Plate 12. A Wilks model 41 vapour phase GLC-IR accessory shown in operation in a spectrometer. (See p. 80.)

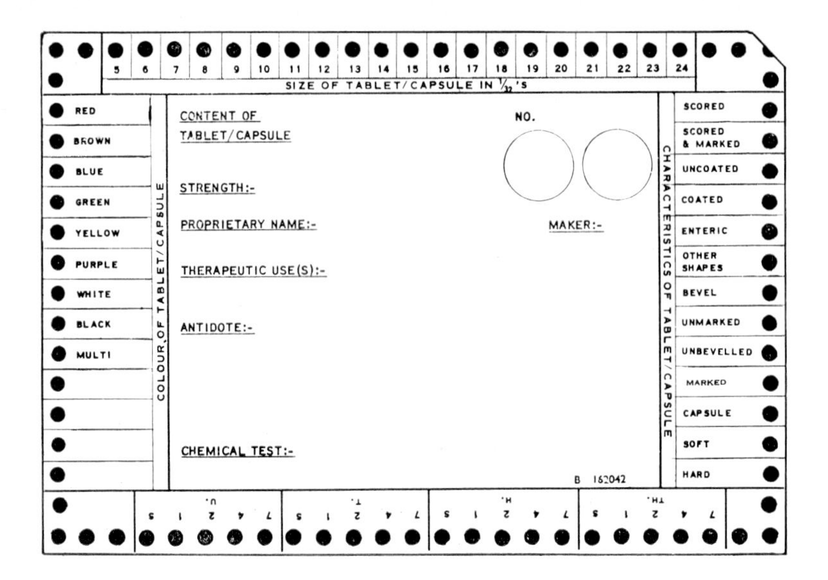

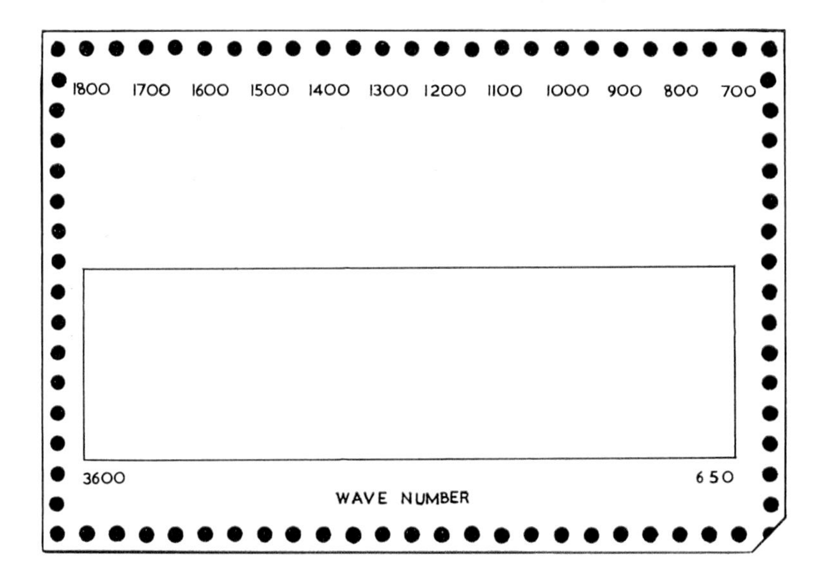

PLATE 13. Tablet identification card of McArdle and Skew. Reverse of the card (lower illustration) used for an infra-red retrieval scheme and DMS-sized spectrum. (Illustration is two-thirds the actual size.) (See p. 179.)

3

ANOMALOUS SPECTRA

Good and quickly seldom meet.
PROVERB

Some of the more common reasons for anomalous spectra have been mentioned in previous sections, a few being discussed in passing. Not all the problems are insoluble, although some cause more trouble than others and need an element of compromise to maintain consistent results. For the most part, difficulties arise in the treatment a sample receives before it is presented to the spectrometer and comparisons between two laboratories or workers are not as easy as comparisons made by the same worker, Five main causes of divergent spectra can be identified and are given in reverse order of probability: instrumental parameters, interference, sample variations, sample preparation, and self-emission.

INSTRUMENTAL PARAMETERS

Modern mass-produced spectrometers designed for routine work have a minimum of possible adjustments, and direct comparisons between different makes of instrument are to be expected on the basis of their similar specifications. This is particularly true if spectra are compared on the same scale as published spectra, such as DMS cards. Differences in dispersion and resolution tend to be smoothed out in the reduction process and only become obtrusive if an original at full size is compared with the reduction. Provided comparisons are made between spectra taken on similar instruments or, preferably, on the same type, gross differences in dispersion and resolution should be rare.

Occasionally, an instrument is badly maintained or installed and becomes out of balance. This condition should be instantly recognized by the water and carbon dioxide bands observed during a blank run. However, this routine check may not be carried out if the spectrometer is in constant demand and the imbalance gets progressively worse without being noticed. Water bands are to be expected from humidity gradients but a positive or negative band around 2350 and 670 cm^{-1} due to carbon dioxide is not logical.

INTERFERENCE

Spurious bands not observed in a re-run are invariably caused by electronic noise or mains fluctuations, and are fortunately a rare phenomenon in a well-maintained laboratory. Regrettably, interference from impurities introduced into the sample occurs more often. In the absence of a duplicate, it may be impossible to decide which of the factors discussed in this section are responsible for the interference, although some clues can be gleaned from the nature of the extra bands. Wholesale swamping by unknown peaks is normally due to interference carried through the clean-up from the sample, or the reagents, and can be identified from the appropriate blank. The presence of a small number of extra bands may be due to the fortuitous contamination by one of the substances given in Table 3.1. A more comprehensive list is given by Launer.[105]

Published spectra are normally acceptable as true records taken from pure material, but there are always exceptions and evidence of impurities may be observed. This was more frequent in some of the earlier atlases before chromatographic and other criteria of purity were adopted and is not unknown even today. Critical appraisal of spectra from different sources is frequently of value and will assist in deciding who has produced the spurious spectrum as a result of impurities.

Interference stems mainly from four sources which are in order of importance: the sample itself, the extraction method, reagents and apparatus, and the worker.

TABLE 3.1

Substance	Approx. wavenumber	Source
Esters	1760 to 1690	Plasticisers from plastics
Hydrocarbon	2950 1450 1370 720 (weak)	Lubricants, tap grease, Vaseline . . .
Carbonate	1430	Rare impurity in KBr*
Nitrate	1380	Impurity in KBr*
Sulphate	1110	Occasional impurity in KBr*
Silicone	1260	Silicone grease
Polythene	730 720	Abraded containers or solubility in solvents
Polystyrene	695	Solubility in solvents
Chloroform	1210 760 750	Traces left in sample
	790	Vapour in the instrument

* Not observed if blank in reference beam.

THE SAMPLE

Natural variation in samples of a biological origin can supply interference of a similar chemical type to the substance being studied, and due allowance should be made:

1. In interpreting the spectrum,
2. By using a blank sample extract in the reference beam,
3. By changing the extraction method, and
4. By a combination of two or all three of these other allowances.

Where the co-extractive is present in variable amounts it is not always practical to make an accurate allowance and the extraction method should be scrutinized.

The Extraction Method

Most materials of forensic interest have been extensively studied and extraction methods developed for a wide range of biological samples. Whilst the acceptance of a method in an official capacity by one of the Associations of Public Analysts in a country is a good guide, it should not be forgotten that a method is valid only for the range of samples recorded and may fail for others; failures are not always recorded. Also, there is always some element of selection for the lowest common denominator rather than the highest common factor in any collaborative study. The value of carrying out one's own collaborative exercise with a 'recommended' method on a range of blank samples before tackling the unknown should never be underestimated. Certain short cuts or modifications can then be justified on the basis of one's own experience. When an unfamiliar sample is encountered requiring micro-techniques, some experience of a proposed method should be gained on controls before the seals are broken—the reputation of another worker is no guarantee when the liberty of an individual may be at stake as the result of the analysis.

Where persistent interference is experienced and is not alleviated by variations on a basic clean-up it may have to be accepted if time does not allow further study. At least by this time it will be recognized for what it is and proper allowance made, and as most extraction methods now involve some form of chromatography, an element of specificity is built into the infra-red confirmation which may be based on one or two characteristic bands only.

Reagents and Apparatus

Reagents and apparatus are rarely a problem if they are suitably selected. Redistillation of solvents is a normal precaution where wholesale evaporations are involved (some American methods are very fond of this) and appropriate selection of other reagents and apparatus is advisable. AR reagents in some situations may be no purer from the infra-red point of view than commercial solvents stored in bulk and so all sources are worth investigating. The most likely source of interference from reagents is a fixed phase used in a chromatographic clean-up, particularly if this is the final stage. A well-conditioned gas chromatograph working isothermally rarely causes interference unless the temperature chosen is near the upper limit for the stationary phase. Temperature programming can be unpredictable in stationary phase 'bleed-off', a factor usually eliminated by dual-column working. Due allowance can be made in infra-red examination of collected fractions by collecting a blank fraction from the second column should the circumstances demand it, although a simpler expedient would be to run a spectrum of the stationary phase, and base the identification of an unknown on bands occurring in the windows; GLC stationary phases are often simple molecules with simple spectra. Materials used in thin-layer and column chromatography frequently cause interference owing to a marginal solubility in solvents or adsorbed impurities incompletely removed during manufacture, or occasionally owing to impurities picked up in storage. Some highly purified adsorbents for chromatography are available on the market, while others have well-characterized limits of impurities and give every satisfaction. Difficulties arise mainly from the relatively drastic treatment used to elute the material of interest from the adsorbent, as most published methods use a direct extract for examination. Silica gel is particularly prone to give silicate interference[59] which could be overcome by further purification involving solvent partition with an aqueous solution. The extra time involved need not be excessive even with a drying and evaporation stage.

Glass wash bottles rather than plastic bottles are essential in micro-work, and plastic stoppers and other apparatus should be avoided or at least checked for potential contamination from plasticisers. Some batches of Teflon may be suitable, others not.

The cleaning of glassware should never be undertaken lightly. As a general rule, detergents should never be used on glassware involved in the final stages of an analysis, and only in other apparatus if it has already been shown to be eliminated by the clean-up process. Prior soaking in chromic acid mixture followed by rinsing in distilled water just prior to use is all that is normally necessary if water is involved in the clean-up, otherwise a final wash in the solvent used in that piece of apparatus is needed after drying. Apparatus with moving parts, such as separating funnels, should be selected to function without lubricants other than the solvents used at that stage; apparatus contaminated with silicone grease should be avoided at all costs.

The effects of inapt choice of apparatus have already been noted.[129]

The Worker

No amount of clean apparatus and purified reagents is of use if the worker is scruffy. An attitude of mind as well as technique is involved. A complete analysis involving many hours work can be ruined by an inappropriately placed thumbprint, or a stopper placed on a polished bench. No potential source of contamination should be dismissed as 'impossible' in micro-work.

The only major source of interference left if the others can be discounted is the overall environment of the laboratory.

External Influences

Because of the expense involved only a few laboratories can afford to place all their apparatus in an air-conditioned room and most usually compromise by placing the spectrometer only in such a room. In the centre of a large city, the dusty atmosphere is a severe hazard and is bound to be the source of interference at some time or other if extractions are carried out in an open laboratory. It is part of the routine good housekeeping of a section specializing in micro-work to monitor interference derived from the atmosphere at regular intervals. Typical spectra of extracts, or dust incorporated into the matrix used, can be filed and consulted as the need arises. Smoke smuts and rubber dust are obvious contaminants near roads and are easily collected on a Millipore or equivalent filter attached to an aspirator. Silicates from glassware may be carried through aqueous extractions and give a broad band between 1000 and 1250 cm^{-1} whilst organics derived from deionized water can give bands almost anywhere. These last two types of interference occur more often than they should and can be picked up in routine checks.

SAMPLE VARIATIONS

At first, variations in the sample are difficult to distinguish from impurity interference unless batches of pure material are studied. Whatever method of solid sampling is adopted, the problem of natural variations in the sample form will always occur to some extent because of the variety of solvents and physical parameters used to isolate an unknown. Some organic chemicals can exist in a wide variety of crystalline forms or polymorphs, and each form can have a distinct infra-red spectrum. There may be no obvious correlation between a published spectrum and one's own derived from the same (pure) substance, simply because the method of manufacture is different. A change of solvent used in (say) a final recrystallization, or a different heat treatment at the drying stage, could easily convert crystals into an amorphous mass, or vice versa, or alter the degree of solvent of crystallization, or induce a polymorphic transition. Crystalline/amorphous changes are the most frequent problem encountered, most people being presented with a non-crystallizable fluid at some stage during their working life. Isolating substances as salts keeps amorphic forms to a minimum, and allows simple aqueous solutions to be used for any final crystallization or evaporation, thereby introducing a constant factor and also minimizing polymorphs; maintain a constant technique to reduce possible crystal effects.

Changes in crystal form are sufficiently common to merit special study when a new material is encountered for the first time. At the very least, the spectrum of an extract from the clean-up of choice should be compared with that of the starting material available from one's own reference standard, or the literature. Any significant difference indicates crystal changes, provided artifacts from the clean-up can be discounted. Where time allows, recrystallizing from different solvents will indicate if crystal changes are a problem in cases where this fact is not recorded in the literature. Study of the spectra from such re-extracted or recrystallized materials is best attempted by the film technique in the first instance, to avoid superimposing more potential variables in the process of preparation as discussed next.

SAMPLE PREPARATION

If infra-red spectra are used infrequently, and then only to 'confirm' other evidence, inadequate experience of the importance of sample preparation and its effects can lead to erroneous spectra and conclusions. This subsection is concerned with factors altering the diagnostic and characteristic features of the true spectrum of a pressed disk.

Of the five major causes of anomalous spectra arising from sample preparation, the first is a physical effect of the disk possible with all samples, and the others are functions of a particular sample involving fundamental changes in the sample.

No discussion of these factors as they apply to pressed disks would be complete without reference to the work of Duyckaerts, and the reader is advised to consult his paper[69] for a further mathematical treatment.

Interference Fringes

A correctly made disk can give interference fringes by reflections from its surfaces. These fringes are characterized by bands of approximately equal intensity in regular progression across the whole spectrum and may not be recognized if they are superimposed on a detailed spectrum of moderate intensity. In these cases there is little risk of a fallacious identification because characteristic bands of the unknown will stand out in their correct proportions from the background, and it is only when the required spectrum is of the same order of intensity as the interference fringes that problems arise. A trough of the interference pattern coinciding with a major band of the unknown, or a peak coinciding with a minor band from the unknown, would be quite sufficient to distort the whole spectrum, and over the whole range covered by the spectrometer, several examples of each are possible. At the same time, the interference fringes could be mistaken for part of the required spectrum.

Interference fringes are best identified in the 2500 to 1800 cm^{-1} region which has few strong bands from most organic materials. Fortunately the effects are not always obtrusive in micro-work because it is unusual for pellets to have the requisite degree of surface polish and parallelism, although a poorly made pellet with internal fissures is another source of interference fringes.

If fringes are noticed as a regular feature in standards prepared by a constant technique, this 'new' spectrum could be taken for reference purposes, or due allowance made for the underlying interference pattern. Care is necessary in either of these approaches as subtle changes in pellet parameters can have a profound effect on the interference pattern.

As one's ability to make disks and pellets improves, interference fringes should be increasingly anticipated.

Particle Size

A large single crystal is one, albeit impractical way of presenting a material to the spectrometer. Real-life solid samples range from well-defined crystals to solidified syrups, and can be presented as such by smearing the sample on an inert support. A recognizable spectrum can be obtained in this way, and in the case of a genuinely amorphous solid, it will differ little, if at all, in quality from spectra obtained by any other method. Where there is any vestige of crystallinity, the effects of crystal (particle) size can be quite profound. In the absence of a suspending medium other than air, energy losses due to reflection from the faces of, and dispersion through, minute crystals are appreciable. These losses are minimized by suspending the solid in a medium of comparable refractive index, and can be eliminated entirely by grinding crystals to a size appreciably smaller than the incident radiation wavelengths. A total disintegration of all particles to the desired degree is difficult to achieve in practice, and with a limited range of refractive indexes available from the limited number of suspending media, spectral effects are quite obvious.

1. *Light scatter.* Mild cases of light scatter superimposed on a general background of sample absorption and radiation reflections from supporting disk/mull interfaces pass unnoticed. The characteristic of light scatter due to large particles is a sloping base-line, rising towards 100 per cent transmission at long wavelengths. In severe cases, the slope is not linear, manifesting itself as a curve towards zero transmission between 2000 and 4000 cm^{-1}. This spectral region is not often recorded and the symptoms frequently pass cursory scrutiny because of the more gentle slope between 2000 and 700 cm^{-1}. Close scrutiny of the spectrum will reveal a general broadening of band widths, with some of the relative intensities altered—in the 4000 to 2000 cm^{-1} range, bands have a very squat appearance by compression into the bottom portion of the chart. Slight alleviation of the condition is possible by using the appropriate fogged disk or mull in the reference beam. 'Straightening' the base-line in this way will not alter the band widths which will remain out of proportion and have incorrect relative intensities.

2. *Christiansen filter effect.* This effect is not always seen or recognized. It is a feature often discussed and described, and yet does not appear to concern an amazing number of workers who form the majority of users of infra-red analysis.

For a direct comparison in simple exercises, this attitude is justifiable to some extent and only becomes untenable when comparisons are necessary, either between workers direct or with the literature. Occasional examples of the effect appear in the literature.

The Christiansen filter effect manifests itself as a distortion of both band contour (to an asymmetrical shape) and band intensity ratios. It is frequently accompanied by a shift in the band wavelength at maximum intensity to the side showing most distortion. The effect may only obtrude in certain parts of the spectrum, as refractive indices vary with wavelength, and the two relevant ones may become widely divergent at these points and yet remain fairly close elsewhere. Because the necessary conditions are discrete crystals of finite size in a dispersing medium of divergent refractive index, the obvious way of reducing its occurrence is a reduction in particle size. This is particularly true where the choice of matrix is restricted to one only, KCl, which has the lowest refractive index of the common alkali halides and can be expected to be different from that of an appreciable number of samples examined.

It may not be easy to diagnose which of the two effects of large particle size is dominant in any given spectrum. Simple dispersion/reflection broadens bands in a more or less symmetrical manner, thereby reducing their intensity, whilst other bands due to the crystal itself may be undistorted, enhanced or even emphasized. Whichever is dominant, it is of no consequence, since neither is acceptable, and measurements made on such spectra are of dubious value. The cure is the same for both, namely, further sample grinding. The ultimate perfection may be unobtainable but a near approximation is within the reach of all.

Sample Transformation

A change in crystalline form, or polymorphism, is such a universal feature that it is wisest to assume it to be a universal characteristic of all samples until shown otherwise for the substance under study.

Three interconversions are possible.

1. *Crystalline/amorphous.* Where the solid's spectrum becomes simpler and approximates more to the solution spectra, conversion to an amorphous form should be suspected; the reverse may also occur. The phase change can take place either during the grinding process because of direct frictional heating or the energy of lattice disruption or during disk pressing; also, it can be due to any combination of these factors. Prolonged grinding is a common culprit leading to amorphous forms.

2. *Solvation.* Solvents are frequently an essential constituent of a crystal lattice and help to characterize the substance in other

physical tests. During sample preparation and pressing, the conditions may be sufficient to de-solvate the molecule (die evacuation can assist this process) to yield a totally different spectrum. Pick-up of solvents should be impossible during either process, but the same cannot be said for water, and a hydrated crystal is feasible from a previously anhydrous form. The quantity of moisture present in a matrix does not need to be excessive in order to hydrate the trace amounts of sample used in halide disks.

Confirmation of solvation is necessary where there is a possibility of co-extractive contamination, and is carried out by working backwards through a 'blank' of the purification extraction process used to isolate the compound. In ideal conditions, the situation would have been anticipated from studies of the pure material.

3. *True polymorphism.* This involves a change from one crystalline form to another. It is convenient to regard the two previous interconversions separately, although in cases where partial change in solvation is possible, the borderline between 2 and 3 is hard to define too rigidly.

Substances exhibiting polymorphism usually exist in an amorphic form and/or solvated crystals, but the converse is not necessarily true. Time will decide how far a study is taken in a laboratory. As a minimum, the spectrum in a simple solvent should be compared with a product recrystallized from all the final solvents used in the various extraction procedures anticipated. In the first instance, spectra of the recrystallized product in paraffin oil mull are preferable (to act as a secondary reference point) before attempting spectra in pressed disks. If time does not allow such a study and an unexpected spectrum is noted during an analysis, treating the supposed drug, say, in a pure standardized form, to a complete blank analysis is obligatory before dismissing it from consideration in an identification. Some barbiturates and steroids have been recorded in many different polymorphs, each of which may have a distinctive spectrum.

Polymorphic transitions are induced at any stage during sample preparation. Major changes usually occur during the isolation process, with interconversions between these being caused by grinding, pressing, the matrix and incident radiation. The possibilities are legion if solvation effects are taken into consideration at the same time.

Comparison of spectra from the same disk over a period of time can be revealing. Unstable phase changes, such as a conversion to an amorphous solid, may reverse spontaneously or be triggered off by the matrix lattice, or even be induced by infra-red radiation during the spectral run(s).

Mild grinding may not complete a change leaving a mixture of two forms of variable composition. Some of the fundamental bands will be common to all forms of the compound, and the overall appearance can easily give the impression of impurities, in the absence of prior experience or reference spectra.

The reverse process can occur and is not restricted to pressed disks. Crystal changes occur in a mull through the action of incident radiation or spontaneously over a period of time. Interactions with the mulling liquid are improbable but sufficient material could nucleate on the supporting cell window to alter the spectral character; grinding often converts a solid into an amorphous form which may be metastable and revert to crystals. It may revert to the original crystal lattice or to a completely different (more stable) one. The possibilities are endless although poorly recorded in the literature.

Mulls are seldom stored for any length of time and are used soon after manufacture. In a busy laboratory, there is always a temptation to make a mull at the end of a working period and leave it (say) overnight for transference to the cell and recording in the morning. Crystal changes in mulls are therefore rarely observed, except on such occasions of a delay, when the possibility must always be considered (see pp. 150, 151, and 157).

Baker[9] summarized the major factors determining polymorphism as a source of solid-state anomalies in pressed disks as follows:

Crystal energy of organic phase.

Energy put into grinding sample and matrix.

Lattice energy of matrix.

Relative stability of polymorphs.

Ability of sample to recrystallize in the matrix.

Tolk[191] studied specific points made by Baker and commented on them.

Heating the prepared disk above the melting point of the sample has been one suggestion for overcoming polymorphism problems with barbiturates.[40] A similar approach has been applied to steroids.[84] Constant technique with barbiturates and steroids is obligatory.

Sample/Matrix Interaction

Sample/matrix interaction is mainly a feature of disks and is of little or no consequence when sample and reference standards are prepared under identical conditions. It is of supreme importance where reference spectra are available only from mull or solution, or where reference spectra are unobtainable and a deduction from first principles is necessary.

In order of probability, five interactions are feasible. They need not proceed to completion.

1. *Ion exchange*. Both anion and cation exchange give different organic substances which will possess divergent spectra to some extent. The effect is not limited to organic substances, however, and Meloche and Kalbus[132] record difficulties from anion exchange in studies of inorganic compounds. They note that surface or sample moisture can contribute to the exchange.

Alkaloids and amines are invariably isolated as hydrochlorides during analysis to overcome any risk of volatilization, whereas the article of commerce is often a hydrobromide or one of a number of other salts. Published spectra are normally only available for these commercial salts, and then frequently only as mulls or solutions. Only recently is it becoming routine for hydrochloride salts to be recorded in KCl disks instead of the previously ubiquitous KBr. Study of published spectra for the same salt reveals some interesting differences due to ion exchange. See also Chapter 7, pp. 113ff.

Hydrochloride salts pressed in KBr often show the spectrum of

the hydrobromide salt. Similarly, exchange with KI is possible in the rare instances where it might be used. Sulphates and organic acid salts are particularly prone to exchange yielding a combined spectrum of base hydrochloride and K_2SO_4 (or potassium salt of organic acid). Calcium salts of barbiturates can cation exchange to potassium barbiturate; potassium/sodium exchange has relatively little effect on the spectrum.

Wherever possible, use the matrix containing the ion common to the organic salt to overcome ion exchange.

2. *Adsorption* is essentially a feature of organic acid salts or hydroxylic compounds. Farmer[75] attributes spectral changes to surface adsorption on halides through hydroxyl groups. Tolk[191] commented on this conclusion.

At first sight adsorption may not appear to be an important factor, but many pharmaceutical bases are formulated as maleates or other organic acid salts and a direct confirmation of this material may be required on an 'as received' basis. Small samples may preclude the use of a simple mull and a pressed disk is necessary. Three possibilities then arise. (*a*) The acid anion can adsorb direct, leaving the base cation more or less intact. (*b*) Ion exchange takes place, the acid anion adsorbing as in (*a*); the base cation spectrum is that of the new salt. (*c*) Traces of moisture convert the material to free base and organic acid, both of which have distinctive spectra of their own. The acid moiety can adsorb as monomer or polymer to add an extra variable.

Most of the work on adsorption has been carried out with benzoic acid and may not be directly applicable here where salts of complex organic bases are concerned. The relative dissociation constants must play some part where the acid moiety has been shown to exhibit adsorption effects, and each case must be treated individually.

Bellanato cites similar effects occurring with amine salts.[15]

3. *Mixed crystal formation* and

4. *Complex formation* are more probable with inorganic materials. Examples of 3 and 4 with organic compounds are discussed by Duyckaerts[69] and Tolk.[191]

5. *New compound formation.* Apart from (2 *c*), the four other interactions above involve changes, or permutations, in materials knowingly used. Traces of impurity in the matrix, not contributing a spectrum of their own, could be present in sufficient amount to form a new compound with the sample if this new compound is relatively more stable.[190]

A fifth source of anomalous spectra is not described very often.

Sample/Cell Interaction

This is fortunately a rare phenomenon restricted to mulls and can be regarded as a special case or a sub-division of new compound formation (5 above). It occurs when a sample reacts with the rock salt plates used as the mounting cell. This reaction normally only happens when the sample is damp and ionic.

SELF-EMISSION

Warm samples emit infra-red radiation. The effect on quantitative estimations is obvious and even diagnostic qualitative studies can be upset. With perfectly transparent disks, sample heating is minimal, and it is only where one of the faults itemized on pp. 16-19 occurs that self-emission should be expected. Small differences in temperature are not harmful and are seldom exceeded by the normal operating conditions of commercial spectrometers. However, beam condensing will cause proportionately higher heating (unless a filter is fitted) and this could diminish the overall anticipated gain by sample self-emission. One way of obtaining an infra-red emission spectrum of a material is the reverse process, namely, lowering the ambient temperature of the detector, normally by a considerable amount. The relatively warm sample then emits characteristic radiation, often identical to the absorption spectrum. This may be recorded by introducing a chopper into the beam between the sample and detector.

Spectrometer design determines whether or not this factor is important. As the signal is unmodulated radiation, it will not be

recorded if the spectrometer chopper is placed between the infra-red source and the sample.[116] Check that your spectrometer design complies with this important criterion.

4

ALLIED TECHNIQUES

They also serve who only stand and wait

MILTON

This chapter does not set out to be all embracing, but merely to be a collection of some of the more original and thought provoking ideas which have come to my notice and appear to work for someone. I have added my own comments where I feel it to be necessary.

A considerable proportion of analytical work will be completed without recourse to infra-red spectra and, for simple confirmations, ordinary colour and physical tests suffice. Except in specialized instances, some separation of mixtures is needed, and these separations provide an extra verification of identity. However, none of these normal preliminary tests can provide an absolute proof of identity in quite the same manner as an infra-red spectrum, although a prior chromatographic separation is a particularly powerful combination. Reviews on preliminary tests and isolation techniques in toxicology have been published in the past, and of these, the work by Clarke[38] is comprehensive and obviates the necessity of any repetition here.

Some of the allied techniques described are concerned with a chromatographic partition, and, of these, thin layer chromatography (TLC) is the most important because of its simplicity, low cost and wide application, that is, in terms of the number of people using it, as well as in the classes of compounds studied. Like all analytical methods, these allied techniques are subject to the same vagaries of human nature as infra-red spectrometry, and one must weigh-up the ability of the worker in an unfamiliar field before letting him loose. A conjunction of bad sample isolation and bad infra-red technique is soul-searing.

A few novel methods of isolating drugs in a form suitable for a subsequent infra-red examination are included as these require a chemical transformation and, thereby, add an extra degree of specificity to the analysis.

GAS LIQUID CHROMATOGRAPHY (GLC)

Although GLC is frequently regarded as an absolute analytical method, it should not always be regarded as such when dealing with a complete unknown of biological origin, even when using a selective detector.[55] The only valid parameters from a GLC record are the retention time and the detector response (where this is selective), because a detector response does not prove anything when a crude extract is injected. Nevertheless, the ability of GLC to regurgitate pure compounds, or simple mixtures of them, from the biological material injected is a very powerful adjunct to infra-red spectrometry, which does confirm the identity of each fraction.

The problems of quantitative fraction collecting are considerable and the effort is rarely made, reliance being placed on the GLC assay, based on a detector response. Fortunately, this assay can be made very precise, certainly within any limits to be expected from an instrumental method, and with electronic integration, it can be made as precise as circumstances demand.

Preparative gas chromatographs are seldom applicable to toxicological analysis by virtue of the small sample size, and the manufacturer's range of accessories cannot be used, whilst analytical gas chromatographs are not always specifically designed for fraction collecting, and the efficiency experienced will be low unless special precautions are taken. It can safely be assumed that GLC fractions are either liquids or solids (unless decomposition occurs on the column, in which case the analyst is in trouble) and the method selected will depend to a large extent on which of these physical forms is being collected. In either case, cooling the effluent gases is necessary, and then the major problem is an efficient recovery of a mist or aerosol.

GLC Fraction Collecting

The majority of methods play safe and adopt the principle of a universal 'cull' and trust to luck for a quantitative return. Aerosols of some solids can be quite tricky to trap and liberties should not be taken with any of the following approaches which can, for the most part, be chosen for gathering both liquids and solids. Where appropriate, solids can subsequently be dissolved in a solvent.

As has been mentioned in the chapter on anomalous spectra (p. 58ff.), care is necessary when interpreting spectra for the presence of the co-eluted stationary phase ('column bleed').

1. *Simple cooling*. Provided the cooling is severe enough, or total recovery is not anticipated, simple freezing of the fractions in a vessel of choice is often more than adequate. Problems arising when transferring fractions to infra-red micro-cells are avoided by combining vessel and cell in some manner. Kendall[98] centrifuged liquids from one vessel to the other, and this is the *modus operandi* advocated for Extrocells;[72] applications for Extrocells, including quantitative aspects, are discussed by Daniels.[53] Blake[17] collected fractions in a simple 'V'-shaped capillary made from glass and transferred them to micro-cells.

Direct application of the collecting vessel as a micro cell has been very common, and of the many references, the work of Molnar and Yarborough[145] and that of Chang[25] have been selected as two typical examples of the use of polythene capillary tubing. By far the most elegant collecting method in this category is the combination of the Peltier effect for cooling and direct ATR examination in the same cell (Plate 11).[207] Sensitivity may be relatively low since a discrete surface film of liquid is required.

It will be noted that all the above examples are concerned with liquid condensates. One simple freezing method is applicable to any type of condensate and is perhaps too simple to appeal to the average complex-orientated mind. Fales and his co-workers[74] were concerned with relatively large-scale preparations of substances of high molecular weight, isolated as solids, and this may account for the poor acknowledgment they have received for their method. Their collecting vessel was a glass 'U'-tube chilled in liquid nitrogen. By using argon as a carrier gas, they obtained excellent results (usually quantitative recovery), which they attributed to the 'rain' effect caused by the condensation of argon. Liquid argon was permitted to evaporate, thus leaving each fraction as a crystalline deposit, which, on further warming to room temperature, will revert to a liquid for low melting point chemicals. The microcrystalline solid or liquid puddle is suitable for infra-red spectrometry. The high cost of argon gas may militate against its universal use as a carrier gas, but in instances where fraction collection is anticipated, the simplicity of Fales' approach[74] has much to commend it.

2. *Trapping*. Apart from Fales' technique, recoveries by simple freezing are generally low and unsatisfactory for solid samples in short supply. The use of electrostatic precipitators referred to by Fales is too elaborate for most workers and certainly lacks finesse. The alternative is to use an inert support, either in the collecting vessel or as a trap in its own right.

(*a*) *Inert support*. The Widmarks[203] studied quantitative aspects, and, therefore, their work is cited as an example. They froze out fractions in an 'F'-shaped tube and found the presence of an inert filling, such as 'Chromosorb W' or glass ballotini, to be essential. Except for low-boiling fractions, liquid air was not necessary, and 'dry ice' gave the same results.

(*b*) *Millipore filter*. General instructions are provided by the manufacturer.[5] More explicit details are described by Hajra and Radin[85] who examined ^{14}C tagged methyl esters, an application outside the present work. Their instructions are sufficient for all practical needs; it is noted that a quantitative recovery is claimed for the filters.

The main advantages of millipore filters are that no elaborate preparation is carried out, that the filters are purchased ready-matched, that a spectrum can be recorded directly (or by ATR), and that, in favourable conditions, a filter can be re-used several times. Filters may be incorporated into KBr pressed disks, although this is a specialist application not discussed in Chapter 1 because of radiation dispersion. Thomas and Dwyer[187] reach a similar conclusion for GLC fraction collecting, and state that the technique is best restricted to low volatility liquids, as good spectra cannot be obtained from crystalline solids owing to the dispersion of the incident beam.

(*c*) *Other supports*. Only one other support appears to have been considered. It is KBr.[106] Some losses are experienced in the collection or the pressing, or both, particularly if the substance is a liquid and a vacuum is applied to the matrix during the pressing. Nevertheless, the principle is sound and should give every satisfaction with solids of low volatility. In this respect, the use of KBr is complementary to Millipore filters.

No attempt seems to have been made to utilize the chemical affinity of nitrogenous bases for silver. Trapping such bases, or their volatile derivatives, in a support of finely powdered AgCl (see pp. 7 and 113), which is then pressed and presented as a disk, would appear to be an obvious extension of the method.

GLC fractions can be handled in any of the ways indicated, or extracted from the vessel with a suitable solvent for a normal solution examination, evaporated and treated as a solid, or any combination. Interference from biological samples is common, and some allowance is generally possible by reference to a control blank sample. Alternatively, further separation by GLC,[203] ordinary solvent partition, or some other method is used. Copier and Van der Maas[45] have adopted some earlier ideas for increasing the specificity of a GLC partition and simplifying the transference to a cell. GLC effluent is condensed in a special trap and subsequently vacuum-deposited on KBr powder. A total analytical time of 6–8 minutes is claimed with a limit of detection at 1 μg.

Another novel method of transferring a sample from the trap to KBr for pressing as a micro-disk is presented by Curry *et al.*[50] Their GLC eluate was collected by direct condensation on a glass surface and immediately dissolved, with washings in 25 μl chloroform. This solution was taken up into a 25 μl 'Hamilton' syringe fitted with a multiple repeating dispenser, and 0·5 μl 'dispensed' to the tip of the needle. About 0·5 mg finely powdered KBr clung to the needle's tip, and by adding further 0·5 μl portions of solution, evaporating each in turn under a table lamp, the whole sample was shifted into the powder. A micro-pellet was pressed in the normal manner. A high recovery is claimed together with the elimination of many possible sources of contamination leading to subsequent high quality spectra.

Curry *et al.* record a substantial loss for materials with an appreciable vapour pressure by this technique, and also a poor recovery from the GLC effluent without drastic cooling with liquid nitrogen or a 'dry ice'/acetone mixture. 10 μg injections of nicotine or amphetamine were found necessary to obtain acceptable spectra. An alternative they gave was converting the amines to a simple less volatile derivative (e.g. Schiff's bases for primary amines) either before injection or on the column, where 1 μg would suffice. The more elegant collection in liquid argon 'rain'[74] or conversion to a derivative after condensing does not appear to have been considered. Hofmann and Ellis's paper[90] is relevant: A stream of anhydrous hydrogen chloride gas from a cylinder passed over the sample converts free amines to hydrochloride salts in a matter of seconds. These salts, being non-volatile, can be stored indefinitely until it is convenient for infra-red spectrometry, and, as they are soluble in chloroform, Curry's transference method is directly applicable. Substituting KCl for KBr would add a final touch.

Some of the collecting and sampling techniques are reviewed in Curry's paper together with details of their GLC conditions giving satisfactory spectra. Most of the references cited there have, therefore, not been employed here.

Direct GLC Examination

All previous comments have been concerned with a two-stage process—first the collection of desirable fractions and then the infra-red spectral analysis of these fractions. Each fraction is an integration of all effluents in the time taken for each trap to fill and fortuitous contamination from some other substance cannot be ruled out. An initial purity check is made by reference to the detector response, from which cues are taken to commence the effluent tapping. The detector response, conventionally shown as a pen recorder trace, is thus the initial lead to which all subsequent answers are related, and may itself only be an integration from many confluent substances. The ability to monitor fractions continually is an extremely valuable feature since it obviates the tedium of collecting and analysing individual fractions.

Although it is possible to monitor GLC effluents directly,[208] to get the best out of the equipment a well-tuned spectrometer with a very fast scan speed is obligatory. The combination is beyond the capacity of all but the most advanced (and well-endowed) laboratories. Two papers by Low[114,115] are concerned with this topic.

Plate 12 shows a commercial infra-red attachment for GLC effluent analysis.

Low's multiple scan interference spectrometry is even more beyond the average laboratory, but it does have the double merits of speed and sensitivity, and can be adapted to gas stream analysis in general. A rapid scan between 2500 and 250 cm^{-1} takes about 1 sec, and this information is stored on tape. It is, therefore, in a convenient form for direct feeding into any of the more advanced data retrieval systems for direct reading or comparisons with previous runs. Interference can be eliminated at the same time.

The future for this technique would appear to be assured since it is a serious contender for the honours currently held by a GLC/mass spectrometer combination in the type, and sheer volume, of information churned out. Infra-red (IR) is not so upset as mass spectrometry (MS) by contamination from other materials of similar retention time or by GLC substrates. The days of a routine combined GLC-IR-MS apparatus giving simultaneous data on the same micro fractions do not seem to be so far off. Infra-red analysis in the gas phase is not subject to the same plethora of sampling errors as occurs in other phases, and has the advantage of ingratiating the theorists.

THIN LAYER CHROMATOGRAPHY (TLC)

For the purpose of this section, paper and column chromatography will be considered together with thin layer chromatography as merely separate aspects of the same topic. Paper chromatography has tended to fall into disfavour, while at the other end of the scale, ample material tends to be available from column chromatography, thus avoiding a host of handling problems. Electrophoresis is widely practised, but is mainly applied to the separation of biological materials, which are outside the present scope. The aqueous substrate is an inconvenience, but no more than this if ATR sampling is adopted.

Nearly all infra-red studies of TLC fractions do not directly examine a plate or powder scraped from it, rather an extract which may be retained and examined as a solution or evaporated and treated by a solid-sampling method. The extract is made with a variety of solvents, chosen either for their suitability as infra-red solvents or for their ability to dissolve the resolved material from the TLC fixed phase. In nearly all cases, the solvent will co-extract some interference from the fixed phase and some allowance is necessary.

Applications are published from time to time in a number of British and American journals. Two less familiar papers have been selected to illustrate the general principles of a combined TLC–infra-red analysis. Fiori and Marigio[78] were mainly concerned with an improvement in the specificity associated with the TLC separation, and proposed modifications to standard separation procedures available up to 1964. Infra-red (and ultra-violet) confirmation were treated more as an adjunct, not the primary confirmation of identity. Nevertheless, their ideas on the improvement of specificity by TLC and paper chromatography are of general interest. In contrast, Goenechea[83] was primarily interested in TLC as a simple separation for a subsequent infra-red confirmation, and consequently treats it as a convenient tool. Conditions for the analysis of various narcotics and soporifics on a preparative scale are outlined, and the same principles can be applied to normal chromatography where ordinary spots instead of streaks are involved. The paper is worth consulting for the method of carrying out the isolation of fractions from the plates and for the spectra cited as examples.

TLC Fraction Collecting

Chromatography relies on the absorptive capacity of a substrate with respect to the solubility in a solvent or mobile phase, and it follows that quantitative removal from good adsorbents must be tedious and, at times, impossible. A good solvent is called for, and it would be naive to expect one to remove only the material of interest. The reader will recall that contamination from TLC substrates was singled out for specific mention in the section on interference (p. 61), and that the work of de Klein[59] was cited as one example of interference from co-extractives from silica gel. Interference of this type is not a long-term problem provided precautions are taken for adequate 'blank' monitoring; it is the unexpected interference which causes all the head-scratching. Visualizing agents do not always reveal all substances on the plate, and if the other invisible spots and smears run concurrently with yours, not only will they be obscured to some extent when the plate is studied by some other visualizing agent, but they will still be there when the band or spot is subjected to some recovery process for infra-red examination.

This is true whatever recovery is selected, but it is not important enough to become an obsession, for the reader will already be aware of all the co-extractive interferences from his preliminary investigations with control samples. Some idea of potential trouble can be gleaned from a trial two-dimensional run, using a totally different solvent in the second dimension. My own preference is for the simplest TLC system giving the degree of resolution for the job in hand.[57]

As far as the techniques for removing samples from the TLC substrate are concerned, the method of choice is of no consequence, provided it has been shown to work with the adopted isolation system. A low recovery, say below 10 per cent, is acceptable if the infra-red spectrum is recognizable, but in TLC, perhaps more than in any other separation, the bewildering choice of adsorbents from numerous manufacturers may mean that a favoured transfer technique is less efficient than expected. It is wise to run a control with all chemicals, including *ALL* the compounds intended to be separated on the plate, and confirm that sufficient is, in fact, removed to be detected.

A few methods have been selected as examples. It must not be assumed that their mention is a guarantee by the authors or me of satisfaction with all classes of compound. The reader himself should confirm the methods under his own conditions.

1. *Simple elution.* In column chromatography there is no immediate problem in setting-up special apparatus since elution is merely continued, with the appropriate solvent, until the desired fraction is collected. The fraction can be subjected to further purification processes or evaporated directly onto a cell window for the infra-red stage. Whether this is in the same solvent (adjusted for bulk, perhaps), as an evaporated film, in another solvent, or by one of the solid-sampling techniques, is of no relevance, but the principle of a simple elution is relevant to other chromatographies.

Paper is the simplest phase to handle. Spots or strips, when cut out, will cram into any crude elution tube, and ought to yield their secrets by judicious application of solvent to the top of the tube.

TLC adsorbents are more of a problem if scraped from the plate, unless the operator has a steady hand and proper tools for the job. Normal micro-spatulas have too blunt an edge and need honing to a razor edge (on the top edge only) before they can be used as a miniature 'fish-slice'. Dissecting needles make excellent tools for the preliminary marking-out and loosening. Few persons have the patience to remove manually spots for elution, and the drudgery can be eliminated if a vacuum line is a standard fitting in the laboratory. The elution tube can be made part of a miniature vacuum cleaner and the spot sucked directly into it. Although a proper sintered-glass filter is the best stop for the fine powder, it is rapidly blocked and becomes increasingly troublesome to clean for re-use, and a simple plug of cotton or glass wool is frequently chosen as an alternative. Cotton wool must be extracted with a solvent before use, and most examples of glass wool irritate the skin. Neither are very retentive for fine powders. Glass fibre filter paper is vastly superior and may be used as it is or as a plug of teased fibres.

If simple elution is chosen, the problem of contamination from apparatus or impurities in, or picked-up by, the solvent is omnipresent. McCoy and Fiebig[128] overcame this by using the elution tube to complete the evaporation of a necessarily small volume of eluate in a capillary tube derived from a Pasteur pipette. After evaporation, the residue was taken-up in 4 to 5 μl chloroform or carbon disulphide, and centrifuged straight into a cavity micro-cell. The authors point out that the technique is only valid for non-volatile materials which are soluble in the solvent to be used for the infra-red measurement. There is no reason why the method should not be modified to cater for other types of sample by changing the solvent and making a mull (p. 26), or one can evaporate on a pressed-disk matrix, and press this, or convert the substance to a salt or derivative on the plate and extract the new item . . .

It is of interest that McCoy and Fiebig did not encounter interference from the adsorbent and yet outlined ways of overcoming it. They also stipulate firm packing of the powder in the elution tube, so that 15 to 30 minutes are required to fill the capillary. They did not find any particular advantage in using the 'vacuum cleaner' device of Millet, Moore and Saeman,[141] and conclude that a choice is a matter of personal preference.

2. *Transfer to KBr.* One of the major objections to simple elution is the necessity of removing the adsorbent from the plate. If the analysis is to be completed in solution, there is little alternative to physically removing the located area as most of the eluents are not convenient infra-red solvents and are difficult to transfer to a micro-cell except by a multiple stage process. However, if the ultimate phase is a pressed disk, the sample may be speedily transferred to the matrix without prior spot removal, that is, *in situ.*

All these direct transfer ideas require good chromatographic resolution with some virgin space around the spot of interest.

Rice[162] outlined his TLC spots in a 'teardrop' shape, cleaned the area around them, and placed a line approximately 2 mm wide by 6 mm long of powdered KBr in contact with the sharp end of the teardrop. He added solvent dropwise from a syringe until the sample was eluted into the powder, broke the point of contact between the adsorbent and powder, and allowed the solvent to evaporate. Between $\frac{1}{4}$ and $\frac{1}{3}$ of the far end of the powder pile was pressed into a micro-pellet for infra-red study. Transfer times are of the order of 2 minutes, considerably shorter than for an elution.

Provided time can be spared, some of the guesswork can be taken out of the movement into KBr by using a Wick-Stick. Wick-Sticks are compressed rods of powdered KBr which concentrate samples by capillary action replenishing solvent evaporated from a pointed end.[202] Good results are claimed,[80] and although half a day's evaporation is needed for most solvents at room temperature, the process can be completed in about an hour at 10 to 20°C below the boiling point of the liquid. Either way, the evaporation may be left unattended and is not very critical.

Wick-Sticks are versatile and can be used directly on some of the 'Instant' TLC systems by simply chopping out the right bit with scissors and dropping it into the evaporation vial. The only limitations I can see to their widespread use are the cost, uncertainty of recovery with an unknown, and the use of KBr. None of these factors are of a sufficient magnitude to condemn the idea. The cost is only relative, and as stated[80] '. . . the operator . . . is free to perform other duties', which is more than can be said for nearly every other idea.

No doubt workers outside America will devise their own versions and try the effectiveness of other compressed powders made to a different specification.

Direct TLC Examination

Elution and transfer to potassium bromide are just two versions of a basic theme, namely, a change of medium from the solid adsorbent to an infra-red transparent medium (or solid film). Few workers have been brave enough to attempt a direct examination of a chromatogram.

There is a patented suggestion of sufficient originality for special mention even though it has not appeared as a commercial venture.

Wilks described a combined ATR-TLC system in 1966:[204] 'An ATR plate is ground optically flat on one side and undulating on the other. The flat surface is coated with a TLC medium and a chromatogram developed in the normal way. A spectrum may be determined by scanning narrow bands of the plate.'

The only objections immediately obvious are the possibility of diagnostic bands being swamped by the adsorbent's bands, and the high capital cost of plates which will be locked-up in relatively few determinations over a period of time. Provided these objections can be overcome, the value of a direct, non-destructive determination for micro-samples is incalculable.

ION EXCHANGE CHROMATOGRAPHY

It is a great pity that ionex techniques are not considered more frequently. Perhaps one of the reasons is the relative glamour of TLC; a preponderance of inorganic applications, a number of them going back more than two decades (too old to be of any use now!) cannot help the attitude either. The main value of ionex is its ability to concentrate ionic materials, sometimes specifically, from a dilute solution containing a wealth of inert solutes. Most drugs and compounds of interest form either ions in aqueous solution or ionic metabolites and are candidates for extraction by ionex rather than by solvent partitions. One of the major problems of solvent extraction from biological samples is a tendency to form emulsions, and many weird and wonderful ideas for breaking an emulsion have been propounded. A good, simple method is given by Stevens, although even this cannot break all emulsions.[181]

Although ion exchange chromatography partition is not proposed as the only answer to the problem, a considerable amount of time and effort devoted to solvent extraction could be avoided if this method were used more often. Its main advantages are the relatively shorter amount of time needed in actual attention, no emulsions to break, and the degree of specificity introduced by a sensible selection of resins. It is also an 'instant' technique in that columns can be prepared and left ready for immediate use, and may be regenerated, many times, during slack periods, or by supporting staff.

The extraction does not need to be by conventional column chromatography, with resin beads, since the beads may be gently agitated in a flask and filtered off after a fixed time. Substitution of resin sheets for beads greatly simplifies handling problems. This is the basis of a routine urine check in America.[64,65,95] Once all the bothersome protein has been removed in this way, the resin can be worked-up by any method you like. Protein-bound drugs and metabolites need prior treatment and are a bane whichever clean-up is employed, and are not peculiar to ionex.

Ionex chromatography is not restricted to isolations from biological samples and becomes a particularly powerful and elegant technique when applied to pharmaceutical preparations. Simple solvent extraction, especially from paediatric compositions, is complicated by co-extracted colouring matter and essential oils, and multiple extractions and back-washings are needed before the end product can be submitted to the spectrometer, unless data are already incorporated in the retrieval scheme. Even amines, alkaloids, and other bases restricted by the Drugs Acts, which are conveniently removed by a direct chloroform extraction or other methods discussed in subsequent sections, are subject to interference, without further separation, when essential oils are present. Similarly, barbiturates and acidic drugs are not always free from non-ionic interference in a simple one-stage solvent shake, and some acidic drugs are not very soluble in organic solvents. They may be removed quantitatively on a strong anion-exchange resin. Blake and Siegal[18] describe one such application to a phenobarbitone elixir. They complete their analysis by nonaqueous titration; it could just as easily be extracted from the simple solvent mixture for infra-red study.

This ability of ionex resins to isolate substances quantitatively is a very important consideration. The assay may be completed by evaporation of solutions, and weighing (an absolute method when you know you are dealing with a single compound), titration, or any other method which will leave the substance in a condition suitable for subsequent spectroscopic confirmation of identity. The answer can also be double-checked by titrating the capacity consumed in the resin regeneration.

One note of warning. Fresh resin will 'bleed' organic matter for some considerable time, particularly if solvents are used in the elution, and steps must be taken to allow for this in qualitative identifications. Choose a resin of as low a water regain as possible consistent with your analytical time. If the ionex separation is merely being used as an assay prior to infra-red confirmation, the column must be tested with preparations of known composition until consistent results are obtained. I well recall the initial scattered results in my suggested assay for sodium propionate to the BSI Specification,[61] because an aqueous solution of an organic ion is quite a good solvent in its own right, and will 'flush' from the column organic material which is immovable in inorganic aqueous solutions. Moral—confirm that the column is equilibrated for your particular analysis before regular service. You may be lucky and get away with it first time, but, in most work, luck and infra-red spectrometry do not often go together.

ISOLATION OF ORGANIC BASES

With a few notable exceptions, all the drugs with restricted possession, without a prescription, are organic bases, either as alkaloids or psychotomimetic amines. The exceptions are largely identified by techniques other than infra-red spectrometry; cannabis and opium, for instance, are botanicals presenting their own problems.

The section on smears (p. 94) is concerned with a simple method of isolating these materials from solid-dose preparations for a rapid confirmation of identity. The method is a straight chloroform extraction. The tablet need not be in aqueous solution, but if it is, the solution does not necessarily need to be alkaline. A number of bases are extracted into solvents as ion-pair complexes in quite highly acidic solutions, and some of the classic separation tables are in error in highly concentrated inorganic salt solutions (say, after a neutralization). Transfer of bases into chloroform as an initial step is, therefore, a logical procedure.

Further clean-up will depend on the co-extracted 'unknowns'. Levine[111] described a partition chromatographic follow-up to separate codeine from pyrilamine and phenindamine which relies on this ion-pair formation. The method appears at first to be a little cumbersome and, from a later paper,[68] some progress on the theoretical aspects is noted. A 98 per cent recovery of methylamphetamine in a cold capsule also containing chlorpheniramine, salicylamide, phenacetin, caffeine and ascorbic acid in approximately 1 hour is encouraging. However, mixtures of this type are difficult to assay without some prior knowledge of the contents, and it is the identification of the contents which is the prime concern of forensic analysis, quantitative assays following later. The major stage of detective work in a similar mixture will probably be preceded by TLC, GLC or classic spot tests. If the data retrieval system is working properly, one will already have a spectrum of a chloroform extract of the medicament. Not all preparations are unique in their composition, and a competitor's product may be in the index and give at least some of the clues needed to identify the unknown.

Separating mixtures of bases is not simple except by chromatography. Most of the information will be to hand from prior TLC and GLC and a good 'guestimate' made from it. Levine's partition is one other method of splitting mixtures. For myself, I would rather devote my time to the application of ionex resins in their most valuable mode—the chromatographic separation of ions of very similar dissociation constant.

Provided bases are known to be present singly, or as simple mixtures, they may be recovered as a derivative or complex salt. A simple chloroform extraction may not be appropriate for one reason or another, not the least important being a low concentration of base; 'extending' the trace as a complex is a useful way of improving a gravimetric assay, as well as adding an extra diagnostic pattern to the base's spectrum. References to some of the more common precipitants are scattered in the literature. Do not take the author's or my word for it, try one of them for yourself and see if they have any application.

Tetraphenylborates

Although tetraphenylboron in its sodium salt is a widely used reagent for organic bases, relatively few attempts have been made to try it for the alkaloids and sympathomimetic amines covered by the Drugs Acts. Fewer still have published spectra of the adducts. Two references are of direct relevance. Matta[124] applied the technique to a few alkaloids among other pharmaceuticals, while Sinsheimer[176] included some amphetamines in a study of sympathomimetic amines.

Precipitation is delightfully simple from a buffered solution, and the product is suitable for recrystallization provided one has ample sample together with the time and inclination. Isolation of local anaesthetics[28] and muscle relaxants[29] are just two selected examples of the reagent's use in infra-red spectrometry.

Reineckates

Like the previous reagent, reineckate complexes are familiar but seldom used in the confirmation of restricted materials. The paper by Chatten and Levi[27] is one notable exception in which *d*-amphetamine, *dl*-amphetamine, methylamphetamine and ephedrine are differentiated. Precipitation is also simple, but the method specifies a 3-hour wait.

In addition to references 28 and 29, phenothiazine tranquillizers have been isolated with the reagent.[209]

Alkaloid Precipitants

The classic alkaloid precipitating agents employed in preliminary screening do not appear to have been considered to any extent. Levi in two papers[107,110] studied the spectra of morphine thrown out of solution by three of the regular alkaloid reagents. His spectra are very unimpressive, but the technique could have wider applications with a little more investigation into conditions.

These three types of precipitant would appear to be the most useful because of the inherent simplicity in preparing both the reagents and the end product. For the more adventurous, the following have been described by at least one worker:

8-Chlorotheophylline Salts (Theoclates)

Lamb and Bope[104] recommend the reagent for the narcotics pethidine, levorphanol and metopon.

N-Cyclohexylsulphamate Salts (Cyclamates)

These have been studied in the infra-red region by Campbell and Slater.[24] No restricted bases were involved in this reference which was concerned with antitussives and antihistamines.

Other Reagents

References 29, 27, 209 also cover between them picrates, chloroplatinates, methyliodides and the *p*-nitrobenzoyl and benzenesulphonyl derivatives. The paper by Rich and Chatten[163] is a good summary of the basic methods as applied to local anaesthetics.

ISOLATING BARBITURATES

The biggest single problem with barbiturates is a strong tendency to polymorphism. The isolation method therefore has a considerable bearing on the ultimate spectrum.[41,136] GLC clean-ups are just as prone to the problem unless conditions for the transference to, say, a pressed-disk matrix are rigorously standardized.[45,50]

Decent TLC clean-ups are hampered by the lack of a sensitive visualizing agent for low levels of barbiturates, and an infra-red confirmation is desirable. The sample may be eluted by any of the methods outlined elsewhere in this chapter and subsequently studied in solution or the solid phase.

Few attempts of a serious nature have been made to isolate barbiturates other than by the frontal attack of solvent partition or chromatography,[117] and I am sure the reader has his own favourite. The work of Dole[64] and Jaffe[95] using cation and anion exchange resins respectively are just two attempts to add a touch of originality, and could be adapted for a final infra-red confirmation. Levi has been active in the field of derivative formation for barbiturates as well as for the bases.

If the reader has three days to spare for a preparation, he can try Levi's copper-pyridine-barbiturate complexes.[109] The *p*-nitrobenzyl derivatives are quicker,[26] but both complexes require purification and appreciable amounts of sample.

Perhaps this explains why both ideas have died a natural death.

5

RECOMMENDED ANALYTICAL PROCEDURES FOR PHARMACEUTICALS

Ninety per cent of analytical chemistry is knowing when to stop.

This brief chapter is a summary of the procedures outlined in Chapter 1. It is a procedure believed to give positive results in the shortest possible time provided sufficient effort has been devoted to a comprehensive indexing or retrieval system, in short, it is a system designed for a practising analyst. However, by its stepwise approach, it will often reveal each component of a mixture which can be identified, if needs be, from first principles, and will allow the complete beginner an equal chance of success. Some of the more complex mixtures will be very difficult for the newcomer without access to an archive containing examples of similar mixtures, but, with luck, one component will predominate and can be recognized and provide the key to the other components—some mixtures are more logical than others. In simple exercises involving a confirmation of another analytical method, the interference from other components is not serious if the required component can be seen against the background.

SIMPLE CONFIRMATIONS

The bulk of infra-red analysis is relegated to the confirmation of a result from another method. In many cases this may well be a 'hunch' derived from the analyst's experience or expertise based on a simple observation. In the limiting case, for instance, with tablets, the hunch will be substantiated by taking accurate physical measurements and/or by reference to a tablet collection or atlas. An authentic sample, or at least some prior spectrum, should be available together with details of composition.

Where the prior spectrum gives details of the sample preparation, follow these instructions and compare spectra. In all other cases, be guided by the nature of the sample, in particular, the quantity available.

Ample Sample

Prepare a paraffin oil mull and examine directly. Sample transformations are kept to a minimum and the moisture content, which can be quite variable even in samples in a virginal state, will not facilitate pressed-disk transformations or cause disk opacity.

Small Samples

Mull preparation remains the technique of choice wherever possible because of its simplicity. Where the sample is required for further study, elimination of the paraffin oil before further work can be done may be tedious, and the sample should be examined as a smear or as a pressed disk.

1. *Smear.* If the sample is soluble in chloroform, or some other nonaqueous solvent, dissolve in solvent and evaporate on a rock salt disk. Examine directly and, where needed, scale expand to display at least the minimum number of bands required for the retrieval scheme.

Wash off with further solvent, or alternatively proceed as 2.

2. *Mull.* Micro-mulls are prepared direct by the normal process but may be more conveniently prepared *in situ* from the smear.

Add a very small quantity of paraffin oil to the smear and grind with another rock salt plate until a stiff paste is formed. This presupposes that an ordinary rock salt disk has been used for the smear and not a specially grooved one (p. 36). The mull will rarely be of high quality and will suffer from large energy dispersions at short wavelengths. The operator must be prepared to allow for this in his assessments and must mask his sample adequately and be prepared to compensate the reference beam with a fogged NaCl disk. Poor mull quality is not serious where a simple confirmation only is required and the obvious limitations arising are recognized.

A micro-mull is also conveniently prepared by Szonyi and Craske's method[185] (p. 26) from the solution washed off from the smear. Be prepared for extra bands due to the solvent not being fully evaporated, and also for a complete change in spectrum due to a solution being dispersed in paraffin oil.

3. *Pressed disk.* Micro-samples may be ground with powdered matrix, pressed, and examined as a pressed disk where desired. In normal practice, micro-samples are dissolved in a convenient solvent (which is added to the powdered matrix and allowed to evaporate) and then pressed; a slight modification allows a direct evaporation of the solvent in the grinding vessel before the addition of powdered matrix which is then mixed by further grinding.

In either event, there are considerable advantages in using a solution washed off from a smear examination for the preparation of a pressed disk. Firstly, the answer may be obtained from the smear and thereby render a pressed disk superfluous or, secondly, where the smear is inconclusive, spectral changes arising from the pressing/grinding process may be of vital significance.

It is not necessary to use the full paraphernalia of pressed disks for micro-work. The Wilks Mini-Press is adaptable.[13]

Press a disk of 65–70 mg matrix and punch out the central window only. Grind this roughly, add 5–100 μg sample and mix by further grinding. Return the mixture to the press, placing as much as possible over the cavity and repress. Check for maximum transmission in the spectrometer by using a duplicate press in the reference beam with a similar pellet of 60–70 mg matrix in it. Attenuate the reference beam, if needed, to give the best spectrum. *N.B.* Further adjustment of the reference beam is possible by 'tailoring' the reference pellet (see pp. 15-16).

4. *Solution.* Micro-samples are often eluted from a chromatographic separation as a solution and, provided the solvent is infra-

red compatible, should be examined in solution. Similarly, a solution prepared for a pressed-disk examination or on its way to, or from, a smear examination should be studied. In a large number of cases no further work need be done for a simple confirmation of the identity of single components or simple mixtures. Most of the simple confirmations will not need compensation for the solvent absorption, and this will greatly simplify handling.

Although solution examination is simple, it does not give information derived from crystal bands in such compounds as optical isomers, and will normally require to be backed-up by one of the solid-sampling techniques. To dismiss solution examination entirely would be foolish, particularly as solutions are involved in nearly every micro-technique, and it is the best medium for quantitative measurements.

UNKNOWNS

The principles to be applied for an unknown are the same as for a simple confirmation, the main difference being in the extra care needed at each stage and the necessity for proceeding logically from one stage to the next and making careful notes. Thus, if it is known that a sample is incompletely dried at some stage, or badly treated, this should be noted and an allowance made for it if a unique answer is not derived immediately. An independent checker cannot be expected to know everything going on and may cast doubt on the whole assessment if an obvious interference or fault is not recognized at the time. The fact that a correct answer has been returned may not be enough in the face of a determined cross-examination.

Basic Procedure

1. It is a wise precaution to determine the spectrum of the unknown on an 'as received' basis. Unique answers at this stage are possible for a number of tabletted drugs where these are essentially of pure material. This initial examination should be completed

as a mull since this is less likely to provoke transformations and is completed very rapidly.

This initial examination is the only valid check for changes in sample, but, more important, it is the only logical way of monitoring recoveries at each stage. Some minor, or even major, components may not be very tractable and could be lost with one of the discarded waste products. Retain all washings and materials until all components of interest have been confirmed.

2. Elucidate as much as possible from the general physical properties of the suspect substance. It is a matter of personal preference whether this stage precedes or follows (1), but there seems to be little point in carrying out detailed physical tests if (1) has already given the answer.

The most valuable ancillary tests in this context are the simple visual tests—colour, texture, and markings on medicaments. Odour is often useful. Tasting is a dangerous occupation with the widespread habit of adding lysergide to almost any other medicament. Tablets and capsules are becoming more and more distinctive as markings, shapes and colours are standardized, and most common products are identified, at least tentatively, from reference collections. Perhaps the most valuable feature of these preliminary tests is the negative information provided, namely, that certain lines of approach are precluded or improbable.

3. Crush solids and treat with chloroform.

(*a*) Where the substance is completely soluble, evaporate a portion on a rock salt plate and record a spectrum. Compare with the original's spectrum.

(*b*) Tablets, capsules and solids incompletely soluble in chloroform tend to settle rather slowly in chloroform and require filtering or centrifuging. If time is not important, allow phases to separate in their own time and examine the clear supernatant liquid as in (*a*).
On a routine basis a filter beaker (p. 39) is recommended.

The original's spectrum will indicate the nature of the excipients in medicaments and in cases where no other component is then obvious, a strong smear spectrum is a good indication of the isolation of at least one active ingredient. Co-extracted excipients are relatively few and are recognizable. Consult the data retrieval system. In cases where a positive answer is returned, which *confirms* other data, there is no need to proceed further except to evaporate all the chloroform solution and weigh the residue. This will confirm at least the order of magnitude of the active ingredient present.

4. A nil return at stage 3 is an important criterion and indicates the presence of an acid salt or a base salt other than a simple hydrohalide. This should have been elucidated at 1.

Dissolve the chloroform insoluble residue, or a further portion of the original material, as appropriate, in water. Filter off insoluble matter; centrifuge off if this insoluble matter is to be examined. Acidify the clear solution with dilute HCl acid. An immediate precipitate indicates an organic acid or barbiturate, although this is a general rule with many exceptions, particularly at low levels.

Extract with chloroform. Combine the chloroform extracts (if the process is to be quantitative) and pass them through a short column of granular anhydrous Na_2SO_4 (well washed with purified chloroform!) to remove the water. Reduce the chloroform solution bulk and examine in solution or in a pressed disk, or evaporate and examine as a smear or mull. Only where the substance is expected to be thermally stable, or traces of moisture are not objectionable, should a wet chloroform solution be evaporated by direct heating.

5. The aqueous residue solution from 4 should contain only organic bases present in the original material as a sulphate or organic acid salt together with the occasional amphoteric material. It may contain nothing since a number of organic bases are significantly soluble in chloroform, even from strongly acidic solution. Basify the aqueous solution with a slight excess of dilute ammonia and re-extract with chloroform. Evaporate the chloroform solution in a small evaporating dish after a few drops of concentrated HCl acid have been added. Examine the residue

as a mull or a pressed disk in KCl, or take up again in chloroform and examine as a smear.

Small amounts of ammonia are carried through and can interfere. These may be removed by back-washing the chloroform extracts from the ammoniacal solution with two small portions of water. Where water is objectionable, as in a smear or pressed disk, the chloroform solution may be dried by passing it through a short column of granular anhydrous Na_2SO_4. Before evaporating the chloroform, convert the dissolved free base to the hydrochloride by saturating the solution with hydrogen chloride gas.

6. The aqueous residue contains inorganic and carbohydrate excipients together with ammonium salts. Reference to the original spectrum and classical chemistry will identify these excipients if it is necessary. A few other organic materials will be carried through to this stage, but all the important drugs of forensic interest are sufficiently soluble in chloroform, at one stage or other, to yield a recognizable spectrum.

Simplified Procedure

A high proportion of tablet/capsule analyses are concerned solely with a simple determination of the presence of a restricted substance—a substance whose unauthorized possession is prohibited—and this means in nearly all cases an alkaloid or sympathomimetic amine. Most of the substances concerned are encountered as hydrochlorides, as free bases, or, in a few cases, as neutral compounds (e.g. Pemoline), and are sufficiently soluble in chloroform to yield positive answers from a chloroform smear.

A number of tablets or capsules contain organic acid salts and sulphates and perhaps the largest single group of these encountered contain the sulphates of the optical isomers of amphetamine. The salt is present at about 2¼ per cent in a simple excipient, frequently lactose. Little or no indication of the amphetamine is obtained by a direct spectral analysis of the tablet or from a chloroform smear.

In all cases where an amphetamine sulphate, or only another base salt, is suspected, proceed as follows:

1. *Quantitative assay*

(*a*) Dissolve a weighed portion of the tablet/capsule in a small volume of 2N HCl. Warm if necessary, but do not boil—starch is troublesome if dissolved at this stage.

(*b*) Transfer with water washings to a 25 ml separating funnel and check that acidity is at least pH 1.

(*c*) Extract with 3×10 ml portions of chloroform. Collect these washings if barbiturates and neutral components are of interest. Retain all suspended matter, together with any emulsion, in the separating funnel.

(*d*) Add a slight excess of 2N NH_4OH solution and check that the pH is higher than an NH_4OH/NH_4Cl buffer (pH 11 or more).

(*e*) Extract with 3×10 ml portions of chloroform, retaining all suspended matter and emulsion. Collect the chloroform fractions in another 25-ml separating funnel.

(*f*) Wash the chloroform extract with 2×10 ml water and discard the water washings. This process will also remove traces of emulsion carried through from stage (*e*).

(*g*) Pass the chloroform extract through a short column of granular anhydrous Na_2SO_4 (washed with chloroform) and collect in a tared vessel.

(*h*) Saturate the solution with anhydrous HCl gas and evaporate to dryness at a low temperature.

(*i*) Weigh the product and calculate back to free base or the original salt (which has been identified by classical acid radical tests where the infra-red identification is inconclusive), and examine by the infra-red technique of choice. Some anhydrous hydrochlorides are volatile in chloroform at elevated temperatures[56] and great care is necessary at stage (*h*) unless it has been shown to be safe to raise the temperature.

2. *Qualitative identification*

Proceed as for quantitative assay except that the quantities used need not be precise and only one extraction of 15 ml chloroform need be applied at stage (*e*), and then proceed from (*d*) on p. 101.

When only amphetamines are known to be present, and small quantities of excipient and moisture are not objectionable in the infra-red identification, the extraction is greatly simplified:

(*a*) Dissolve a small portion of crushed tablet/capsule in dilute HCl acid. Solution may be completed in the 25 ml separating funnel to be used next.

(*b*) Add an excess of dilute NH_4OH and extract with one portion of 15 ml chloroform. Retain all emulsion and suspended matter in the funnel.

(*c*) Collect the chloroform extract in another 25 ml separating funnel and shake with 5–10 ml water. This purification stage is optional but is generally a wise precaution to eliminate ammonia and traces of emulsion.

(*d*) Run the chloroform extract into an evaporating basin, add 5 drops of concentrated HCl acid and evaporate to dryness on a water bath.

(*e*) Examine the residue as a chloroform smear, mull, or pressed disk.

(*f*) The evaporated residue is frequently in an amorphous state. It can be induced to crystallize by any of the classical methods of which a simple period of standing is the most convenient, if time consuming. Better results are obtained by adding a further 5 drops of water to the residue and re-evaporating, occasionally stirring.

3. *Resin complexes*

A few commercial products do not contain organic bases as salts but as a resin complex. At least one American company specializes in this formulation for a wide range of compounds including alkaloids and hypnotics, and other companies in Germany and Spain present sympathomimetic amines in this way.

These products are all recognizable by their characteristic appearance (mainly capsules) and the small resin particles present in them (in one Spanish tablet, the resin is powdered).

(*a*) Dissolve the whole tablet or capsule in water and filter off the resin (and insoluble excipients). A small glass-fibre filter paper is preferred.

(*b*) Examine the aqueous solution for drugs by one of the methods given above.

(*c*) Shake the resin residue with 2N NH_4Cl solution for 10–30 minutes and then proceed from stage (*d*) in the quantitative assay at stage (*b*) in the qualitative identification above.

(*d*) A nil return from (*c*) indicates that the resin complex is of an acid. Extract with 2N HCl if identification is needed, and transfer to an organic solvent before proceeding.

The following six spectra illustrate the application of the simple methods described in this chapter.

Fig. 5.1 is a mull spectrum prepared directly from the unknown tablet, and Fig. 5.2 is of a chloroform smear also prepared directly from the tablet—preliminary tests 1 and 3, p. 96ff.

Fig. 5.3 shows a mull spectrum of the product derived from the qualitative identification method of the simplified procedure—method 2, p. 100. The simplified procedure is therefore applicable to complex mixtures, provided some idea of the ingredients is available from the preliminary tests, direct examination and smear. Codeine, which is completely obscured in the direct spectrum, is very easily confirmed.

The remaining spectra are presented for comparison with other components extracted, and with 5.1 and 5.2.

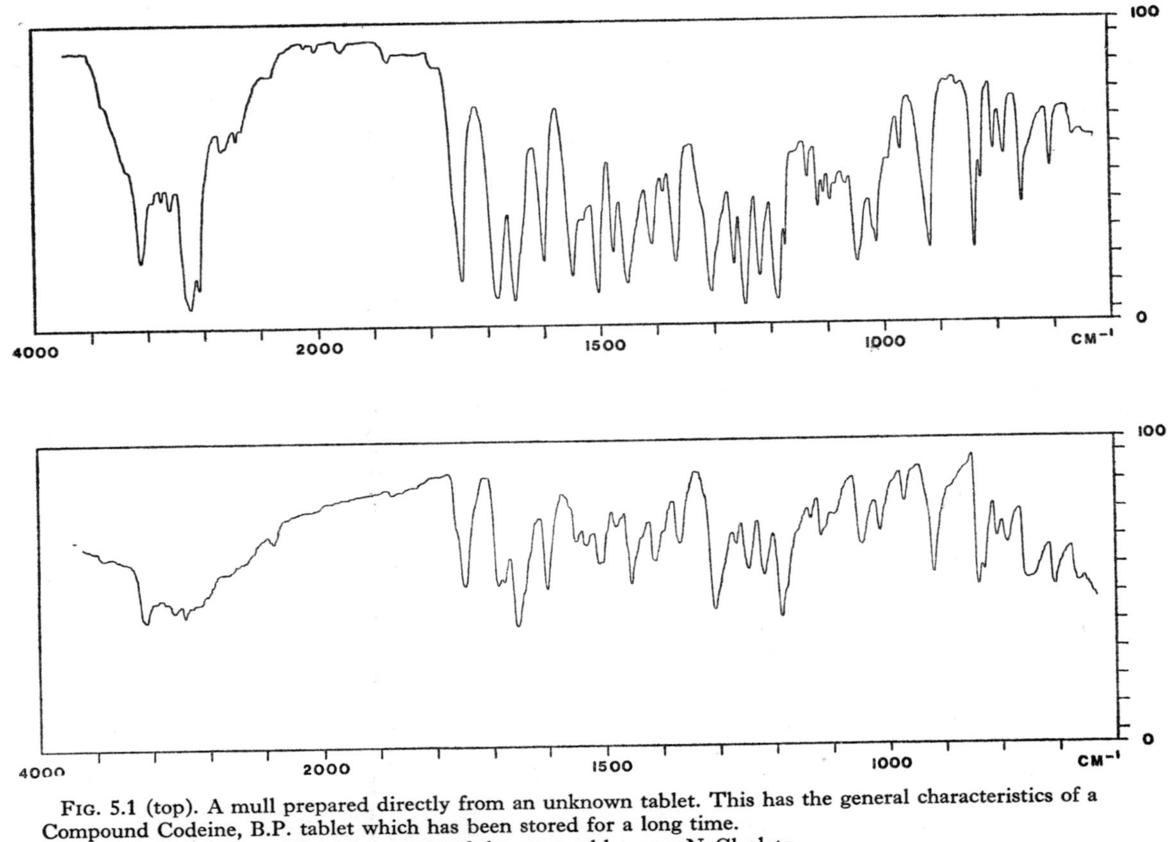

FIG. 5.1 (top). A mull prepared directly from an unknown tablet. This has the general characteristics of a Compound Codeine, B.P. tablet which has been stored for a long time.

FIG. 5.2 (bottom). A chloroform smear of the same tablet on a NaCl plate.

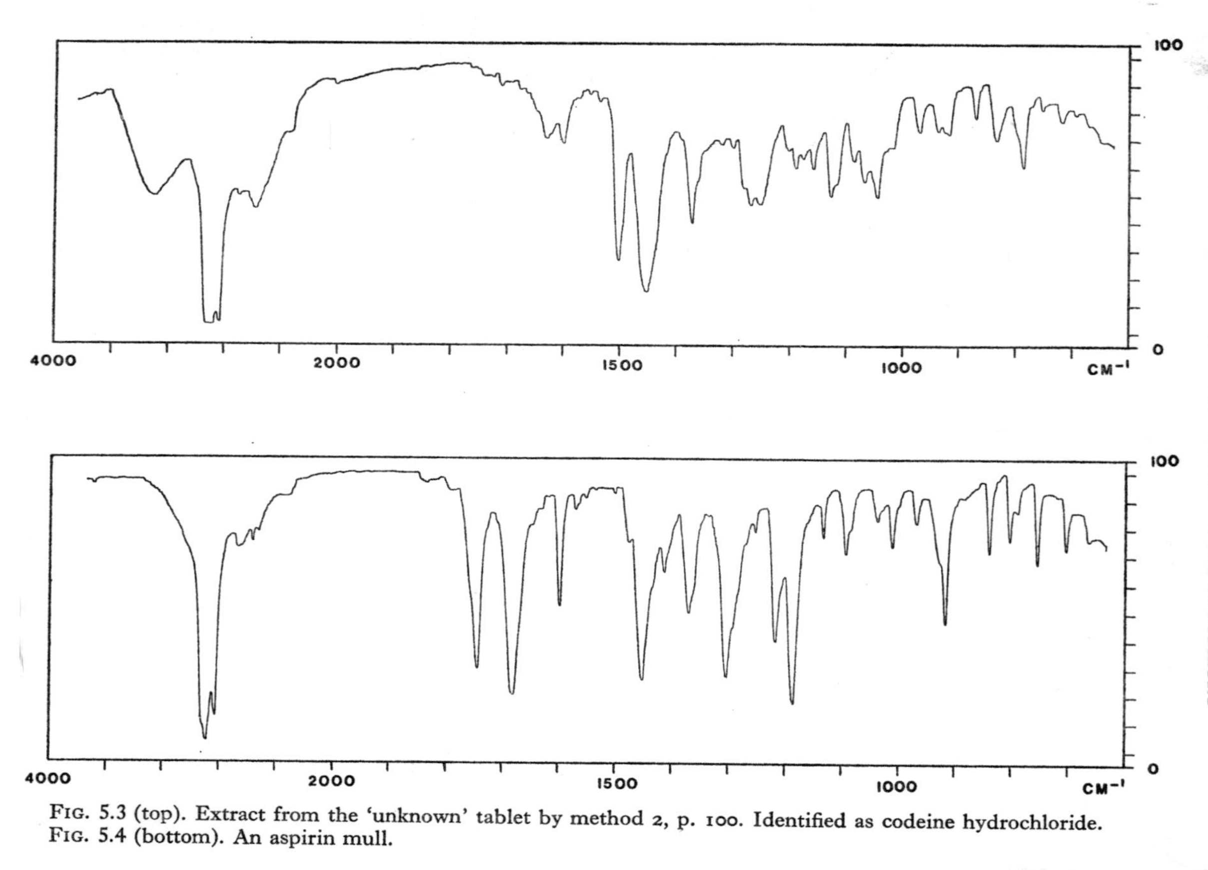

FIG. 5.3 (top). Extract from the 'unknown' tablet by method 2, p. 100. Identified as codeine hydrochloride.
FIG. 5.4 (bottom). An aspirin mull.

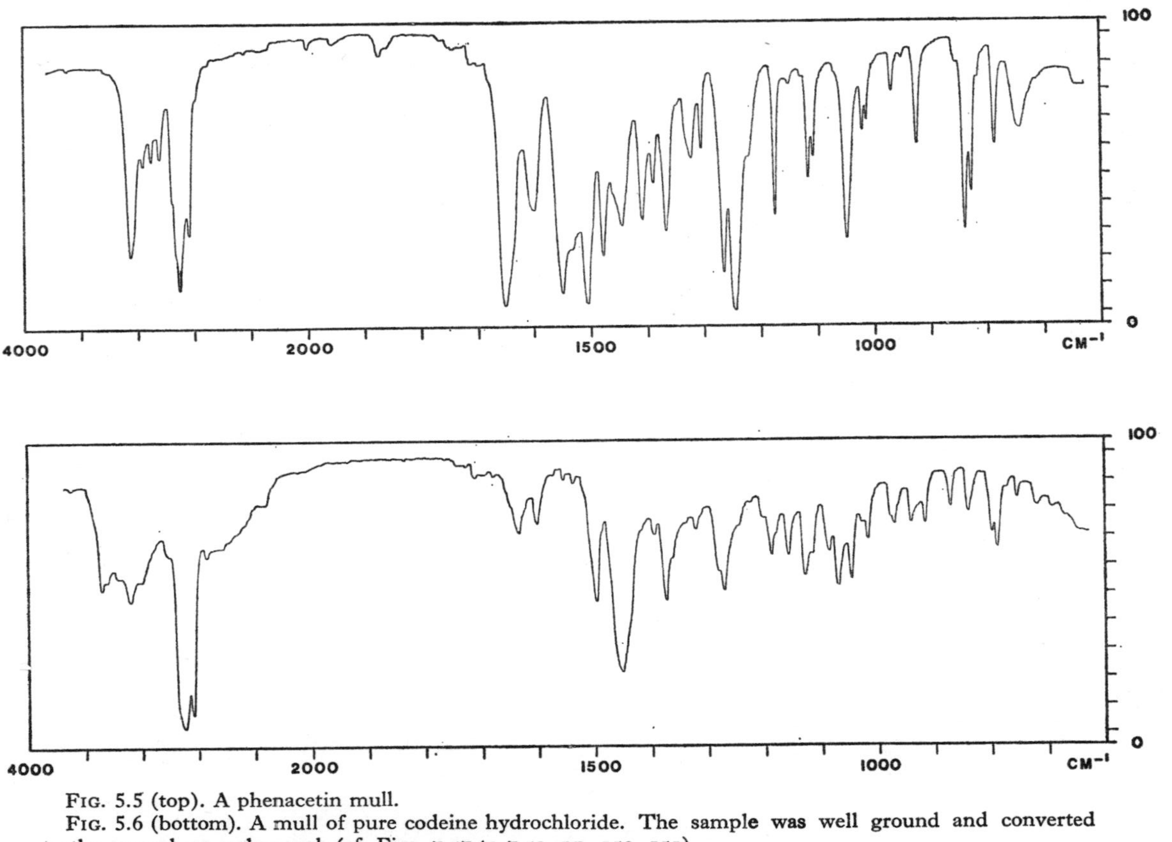

FIG. 5.5 (top). A phenacetin mull.
FIG. 5.6 (bottom). A mull of pure codeine hydrochloride. The sample was well ground and converted to the amorphous polymorph (cf. Figs. 7.57 to 7.59, pp. 150, 151).

6

CHARACTERISTIC ABSORPTIONS OF CERTAIN CLASSES OF DRUGS

> A moment's insight is sometimes worth a life's experience
>
> O. W. HOLMES

A section in Chapter 10 (pp. 195–7) is concerned with band assignment and brief mention is made of the value of diagnostic bands and groups of bands. Recognition of functional groups on a Colthup–Bellamy basis may indicate the presence of particular groups, from which the molecular jig-saw may be constructed, but an extension of this principle to classes of drugs is possible where

1. A pattern of absorption bands is characteristic of a particular combination of groups; or
2. The functional group is only found in certain classes of drugs.

CHARACTERISTIC PATTERNS

Barbiturates

The combination of CO and NH groups in the barbituric acid gives a rise to three bands between 1780 and 1670 cm^{-1}, which may not be completely resolved. Experience with authentic materials will clarify this point. In the solid phase, the usual appearance is of a very strong, broad band near 1700 cm^{-1} (invariably the strongest in the spectrum), with a weaker band near 1760 cm^{-1}. In addition, there is a very broad band near 830 cm^{-1}.

N.B. Some hydantoins give a similar spectrum.

The paper by Manning and O'Brien[119] gives the spectra of several barbiturates, but four of them are salts, although this is not stated, and these spectra are therefore erroneous.

Spectra of the original barbituric acids are given with their derivatives by Chatten and Levi,[26] and Levi and Hubley.[109]

For further details see references 40, 41, 119, 135, and 193.

Corticosteroids

Steroids containing both 20-one groups and 4-ene-3-one or 1,4-diene-3-one (including 21-acetoxy-20-one) usually show three (or more) prominent bands between 1750 and 1610 cm^{-1}, of which the middle one, normally around 1670 cm^{-1}, is usually the strongest.

An incidental, and useful, feature of most steroid spectra is that, after a continuous series of bands from 1470 to 850 cm^{-1}, there are no strong bands below 850 cm^{-1}.

For further details see Mesley.[134]

Sulphonamides

Strong bands near 1145 and 1090 cm^{-1} and a less strong band near 1320 cm^{-1} (not always distinguishable from neighbouring bands) are characteristic of *p*-substituted benzene sulphonamide derivatives—effectively, sulphanilamide derivatives.

Other groups containing the sulphonamide group, e.g. the thiazides, do not have the strong band at 1090 cm^{-1}.

Caution. There is some similarity between these absorptions and the bands at about 1290, 1130 and 1080 cm^{-1} in phthalate esters, a common interference. The absence of a C=O band near 1730 cm^{-1} will rule this out.

For further details see references 34 and 211.

Penicillin Salts

There is a strong band at 1770 cm^{-1} and others at 1680 and 1600 cm^{-1}, the latter being broad and usually the strongest in the spectrum. The 1770 cm^{-1} band, due to the β-lactam, is notable for its high frequency, although some barbiturates also absorb at this frequency.

These remarks do not apply to the free acids; amphoteric penicillins such as ampicillin do comply.

For further details see Wayland and Weiss.[200]

CHARACTERISTIC GROUPS

Amphetamines

The strong bands at about 745 and 700 cm^{-1} are indicative of monosubstituted benzenes in general, but, among the drugs, the only major group in which these bands are the most conspicuous feature are the amphetamines.

For further details see Mesley and Evans.[138]

Phenothiazines

A very strong band at 750–740 cm^{-1} is characteristic of 1,2-disubstituted benzenes. The phenothiazines are one group of drugs in which this band is very prominent.

The triptylines, which would be expected to behave similarly, all show 3 or 4 bands in this region.

For further details see Warren.[196]

Tryptamines

Tryptamine derivatives are another group with a prominent band at, or near, 740 cm^{-1}, provided the tryptamine nucleus is not substituted at the 5- position. Free bases which crystallize are characterized by a remarkable complex absorption extending from 3300 to 2500 cm^{-1}.

For further details see Mesley and Evans.[138]

Lysergide

Two bands at 775 and 745 cm^{-1}, with a strong band near 1615 cm^{-1}, may indicate lysergide (LSD) or some closely related compound.

For further details see references 137 and 192.

The examples chosen above are only intended as a guide, with selected references as a lead to finding typical spectra. There are many exceptions, both of drugs which do not show the anticipated absorptions, and of other compounds which do.

Bad sample preparation can lead to an incorrect diagnosis. The spectrum reproduced in Fig. 6.1 is of a common extract from a tablet subject to abuse. It is soluble in chloroform; this sample has been ground in a typical manner under paraffin oil. The two strong bands near 750 and 700 cm^{-1} could be confused with an amphetamine. Attempt question 13 on p. 209.

References 121, 177, and 188 have general application to the study of amine salts, which are by far the largest single class of drugs of forensic interest. Reference 121 itemizes some of the alkaloid NH frequencies.

Amphetamine studies can be simplified by a knowledge of the possible side chain substitution of benzene (see Potts[155]). Phillips and Mesley[151] have applied a similar approach to the study of unknown compounds. Included in their paper is a study of mescaline analogues and this is also dealt with in a paper by Degon.[58]

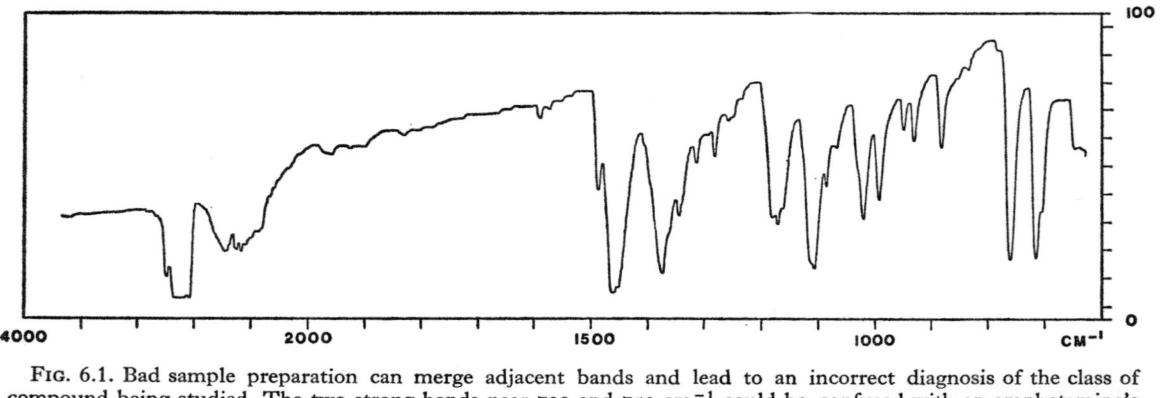

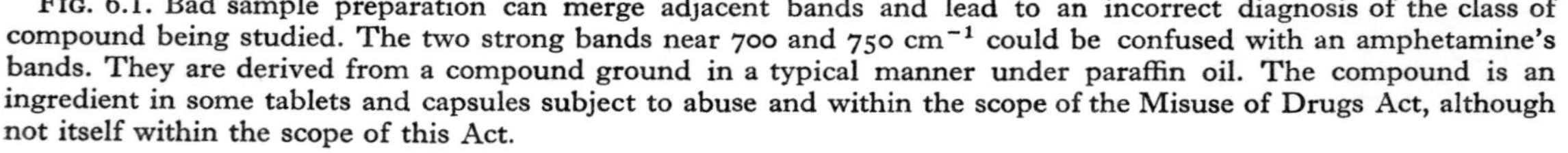

FIG. 6.1. Bad sample preparation can merge adjacent bands and lead to an incorrect diagnosis of the class of compound being studied. The two strong bands near 700 and 750 cm^{-1} could be confused with an amphetamine's bands. They are derived from a compound ground in a typical manner under paraffin oil. The compound is an ingredient in some tablets and capsules subject to abuse and within the scope of the Misuse of Drugs Act, although not itself within the scope of this Act.

7

EXAMPLES

I don't know what effect these . . . have upon
the enemy, but, by God, they terrify me.

DUKE OF WELLINGTON

This chapter gives selected spectra to illustrate the importance of technique and some of the more obvious pitfalls. No solution spectra have been included since these rarely give trouble once the cells have been correctly balanced; much useful identification work can be completed without balanced cells, or even without any compensation for the solvent.

The substance chosen for most of the examples is methylamphetamine (referred to as M in the legends to the spectra). This material is of considerable interest from the legal point of view, both from its topicality and widespread abuse. However, the main reason why it has been chosen is the simple fact that, in the form of a hydrochloride salt, it is a soft crystalline solid, and so should easily produce a good spectrum. It also has a relatively simple spectrum with well separated bands. Its softness is of great advantage for studying direct compression and lamination.

All the pressed-disk spectra have been derived from pellets pressed in a Wilks Mini-Press. This press compares very favourably with more elaborate ones and has the great merit of extreme simplicity. No attempt has been specially made to dry reagents before or during pressing. I have attempted to reproduce these spectra under normal 'worst possible' working conditions. Where water has proved to be important, this has been noted.

Mull spectra have been prepared by prior grinding of no more than 3 mg of sample in a small agate pestle and mortar, and transferring the result to full-sized (25 mm) NaCl plates. There has been *no* masking of the mull in any example and no reduction of the sample beam size or area; the mull more than completely fills the beam with little or no wastage.

All the spectra in this chapter were obtained on an Infrascan, a medium resolution (grating) instrument with the advantages of high scan speed and presentation at simple multiples of DMS chart size. The spectra have been reproduced at DMS size to facilitate measurements or tracing to DMS cards for those wishing to do so.

All spectra, unless otherwise stated, have been attenuated in the reference beam to achieve the best possible spectrum. Also, the chart settings are subject to normal working tolerances, whilst the scan speed has been the fastest possible (4 minutes for a complete scan on the Infrascan). These spectra are, therefore, as near as possible to those produced by the average worker carrying out routine work. Some extra value can be obtained from the examples by trying at least some of the questions on pp. 206–211.

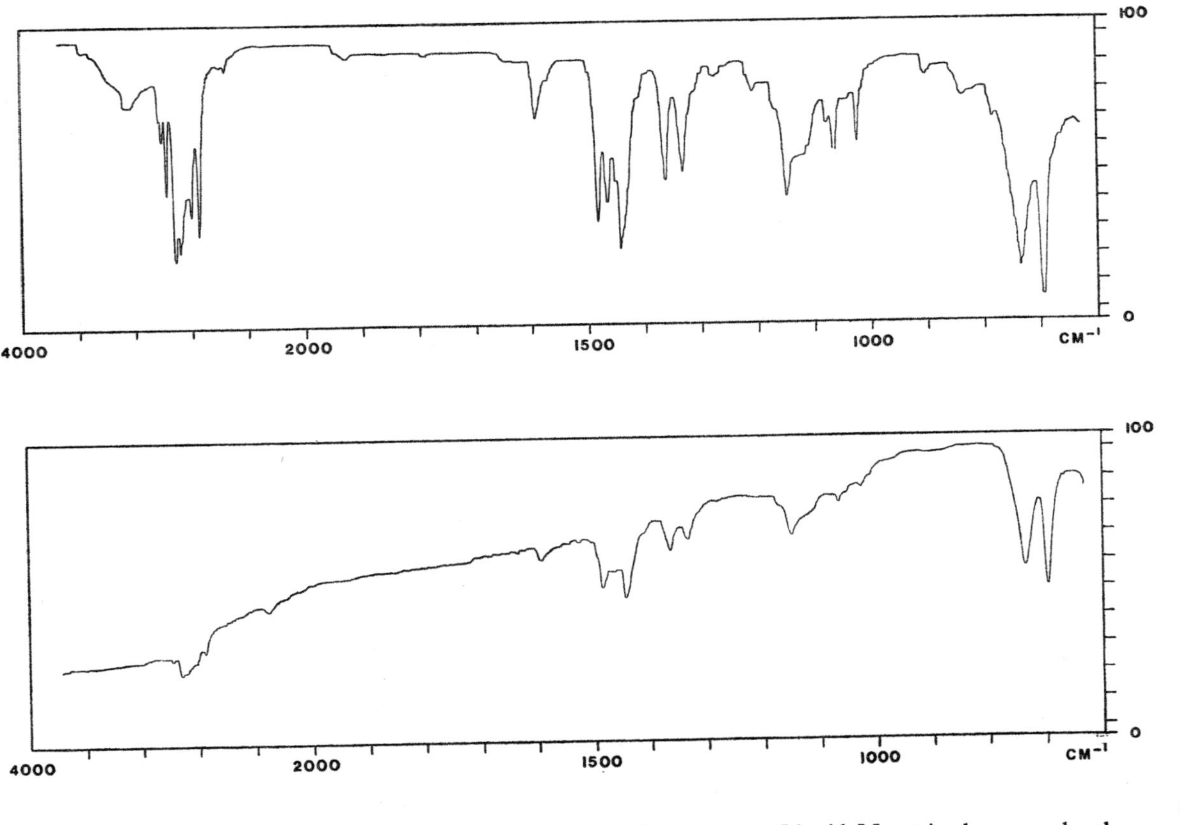

Fig. 7.1 (top). Liquid methylamphetamine (hereinafter referred to as M). Film between two NaCl plates.

Fig. 7.2 (bottom). Liquid M retained on powdered AgCl which has been pressed without further treatment. N.B. No compensation for light scatter.

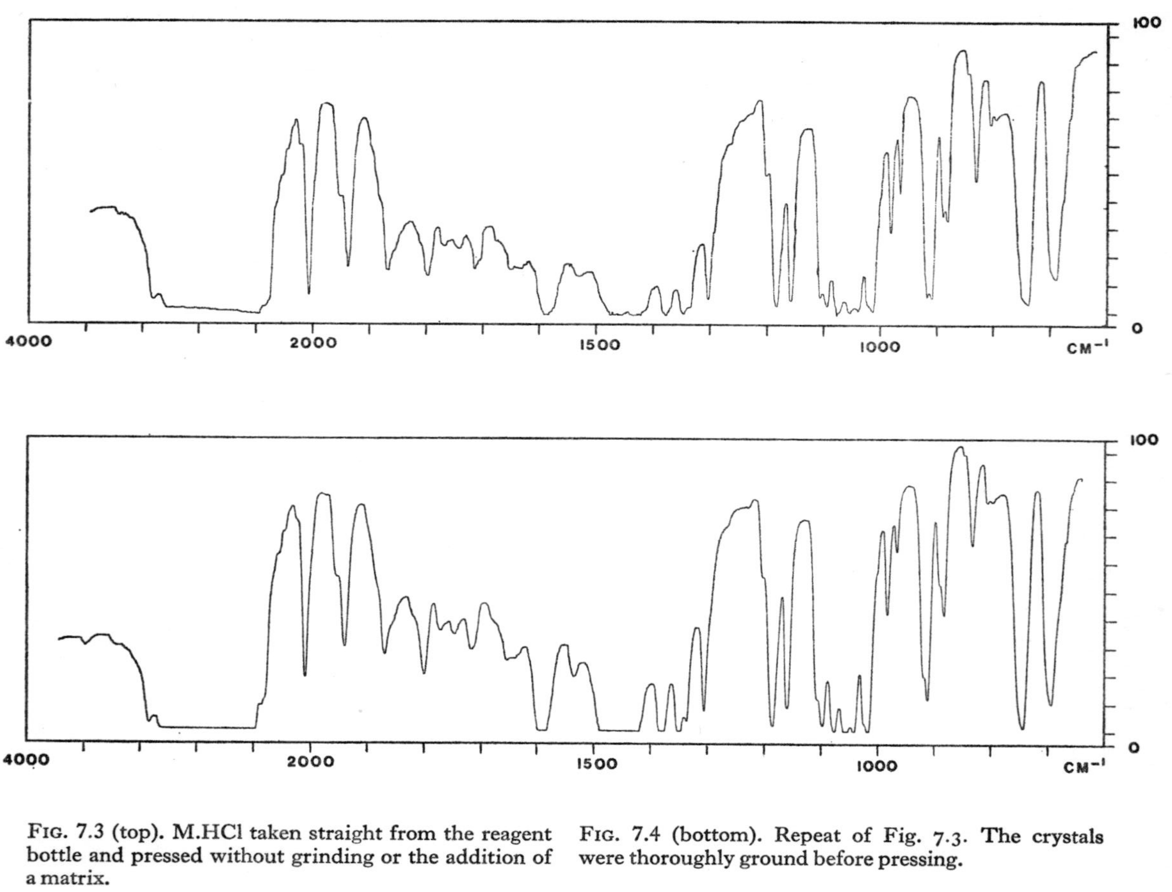

FIG. 7.3 (top). M.HCl taken straight from the reagent bottle and pressed without grinding or the addition of a matrix.

FIG. 7.4 (bottom). Repeat of Fig. 7.3. The crystals were thoroughly ground before pressing.

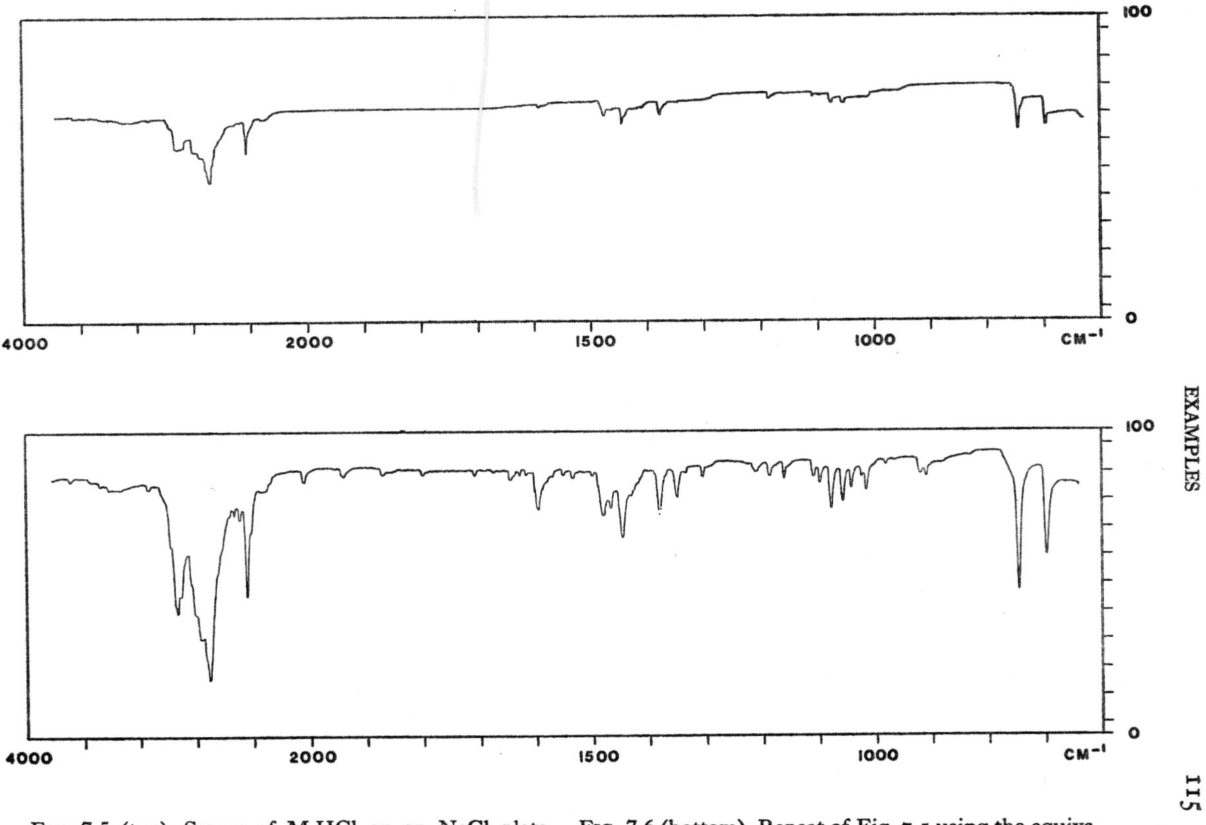

FIG. 7.5 (top). Smear of M.HCl on an NaCl plate. Typical of extract from about a third of a 5 mg tablet to the B.P. specification.

FIG. 7.6 (bottom). Repeat of Fig. 7.5 using the equivalent of a whole 5 mg tablet, and balancing the reference beam with a fogged NaCl disk.

Effect of the Matrix in Pressed Disks (Figs. 7.7—7.28)

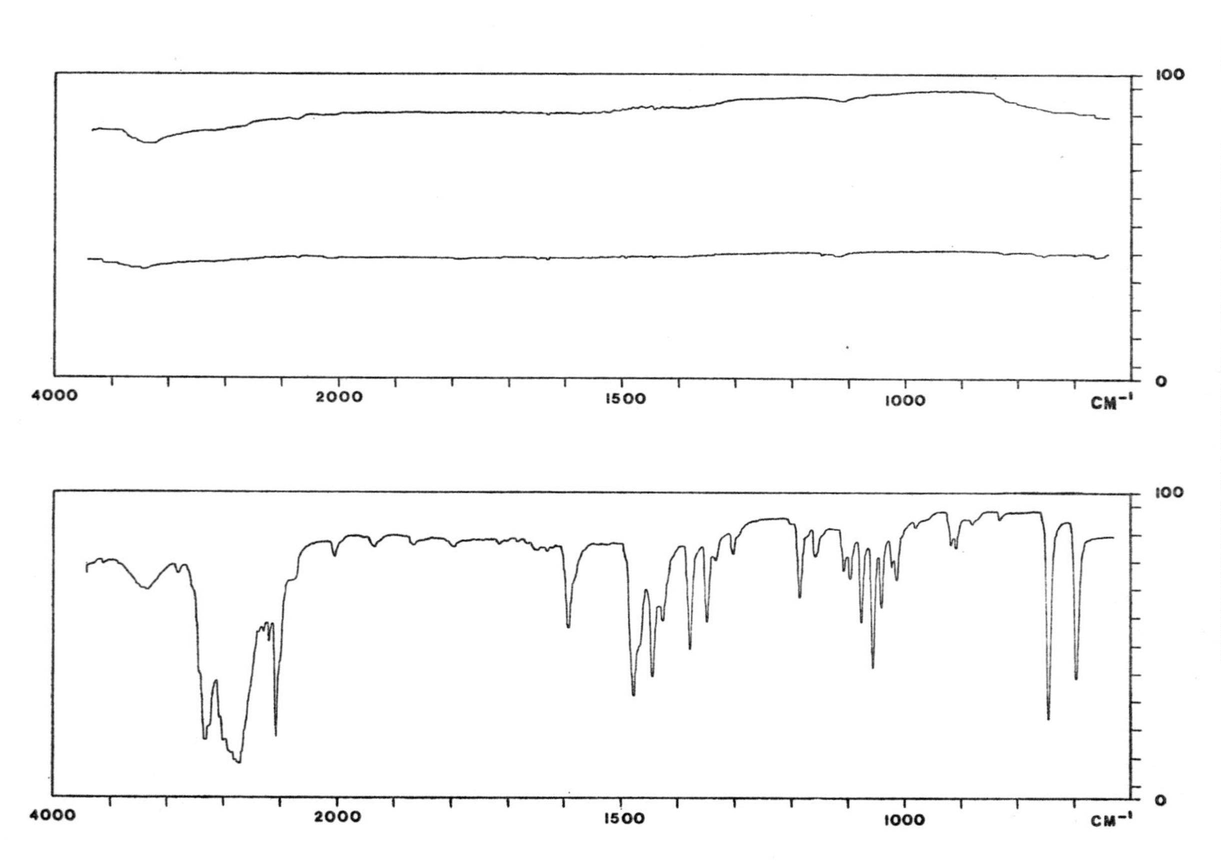

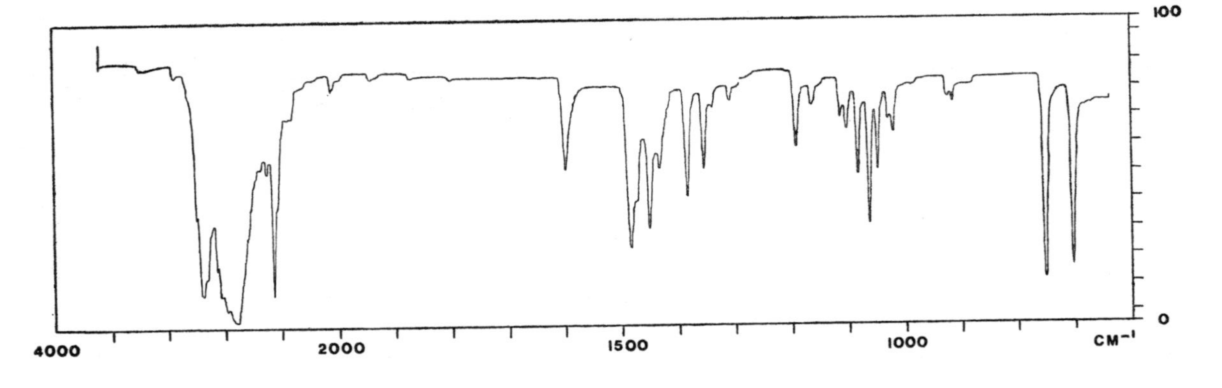

FIG. 7.7 (top). Analytical Reagent grade KCl. Substance dried at 110°C for 2 hours, ground and pressed without further treatment.

Lower tracing is the disk as produced without any reference beam attenuation.

Upper tracing is the same spectrum with reference beam attenuation.

FIG. 7.8 (middle). M.HCl in KCl.

FIG. 7.9 (bottom). The same spectrum as Fig. 7.8, but balanced with a duplicate pellet of KCl in the reference beam. There are noticeable dead-spots in the tracing.

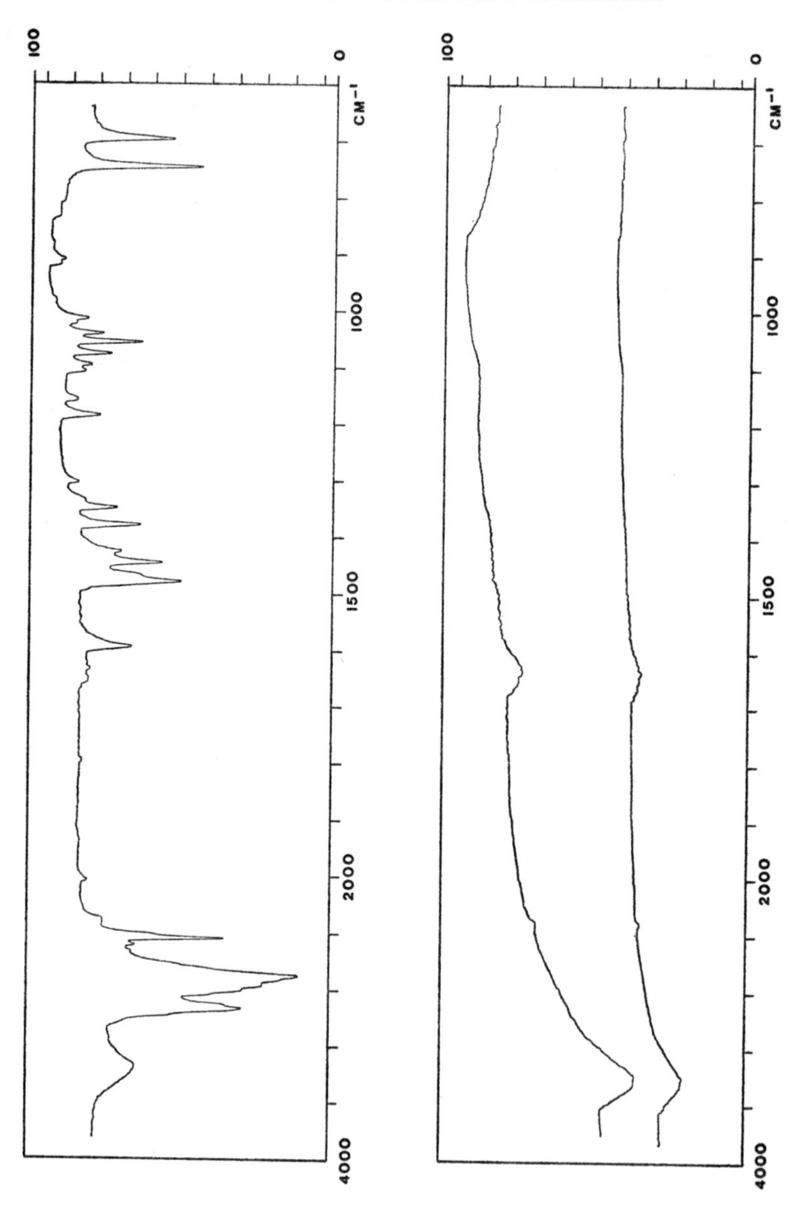
100
0
CM⁻¹
1000
1500
2000
4000
100
0
CM⁻¹
1000
1500
2000
4000

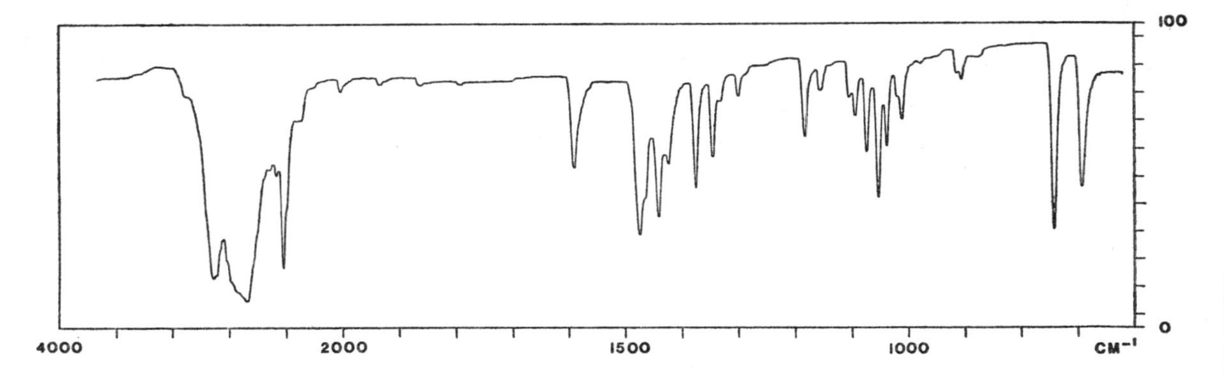

FIG. 7.10 (top). M.HCl in KCl. Sample and small portions of KCl triturated together in a small agate pestle and mortar without undue care and attention. This technique was adopted for a direct comparison with the other substances chosen as a matrix.

FIG. 7.11 (middle). Analytical Reagent grade NaCl. Same conditions as Fig. 7.7. Requires a much higher sintering pressure and is prone to produce opaque disks.

FIG. 7.12 (bottom). M.HCl in NaCl. Reference beam compensated by an opaque NaCl pellet.

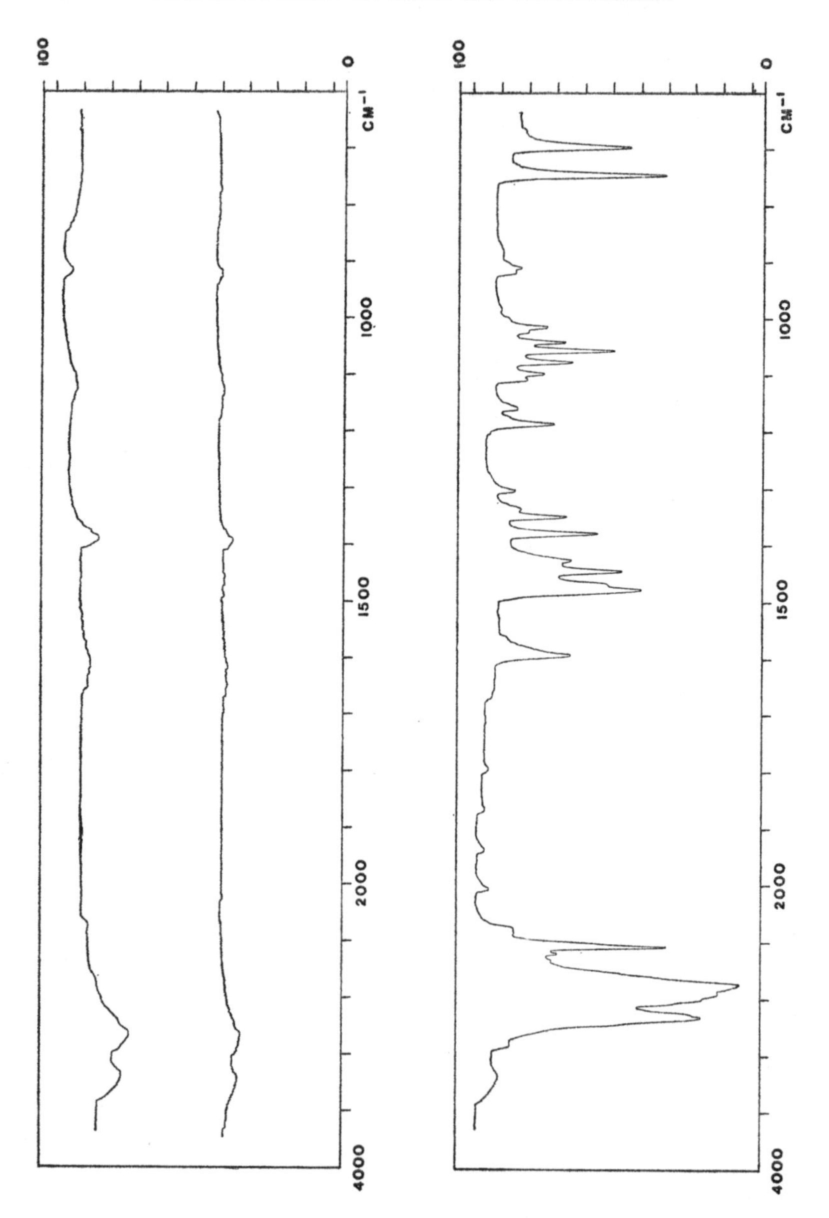
100
0
CM⁻¹
1000
1500
2000
4000
100
0
CM⁻¹
1000
1500
2000
4000

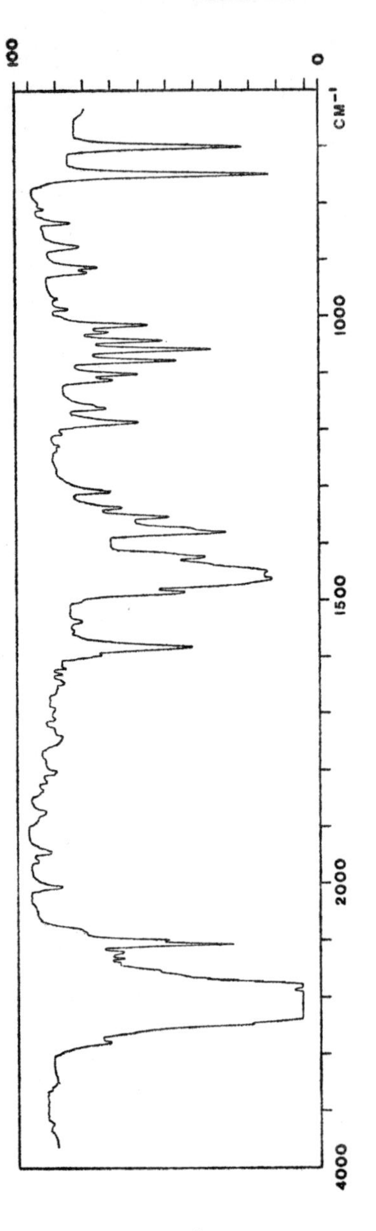

Fig. 7.13 (top). Analytical Reagent grade KBr. Same conditions as Fig. 7.7. Has a tendency to produce opalescent disks.

Fig. 7.14 (middle). M.HCl in KBr. Reference beam compensated by KBr pellet.

Fig. 7.15 (bottom). Mull in paraffin oil of M.HBr for comparison.

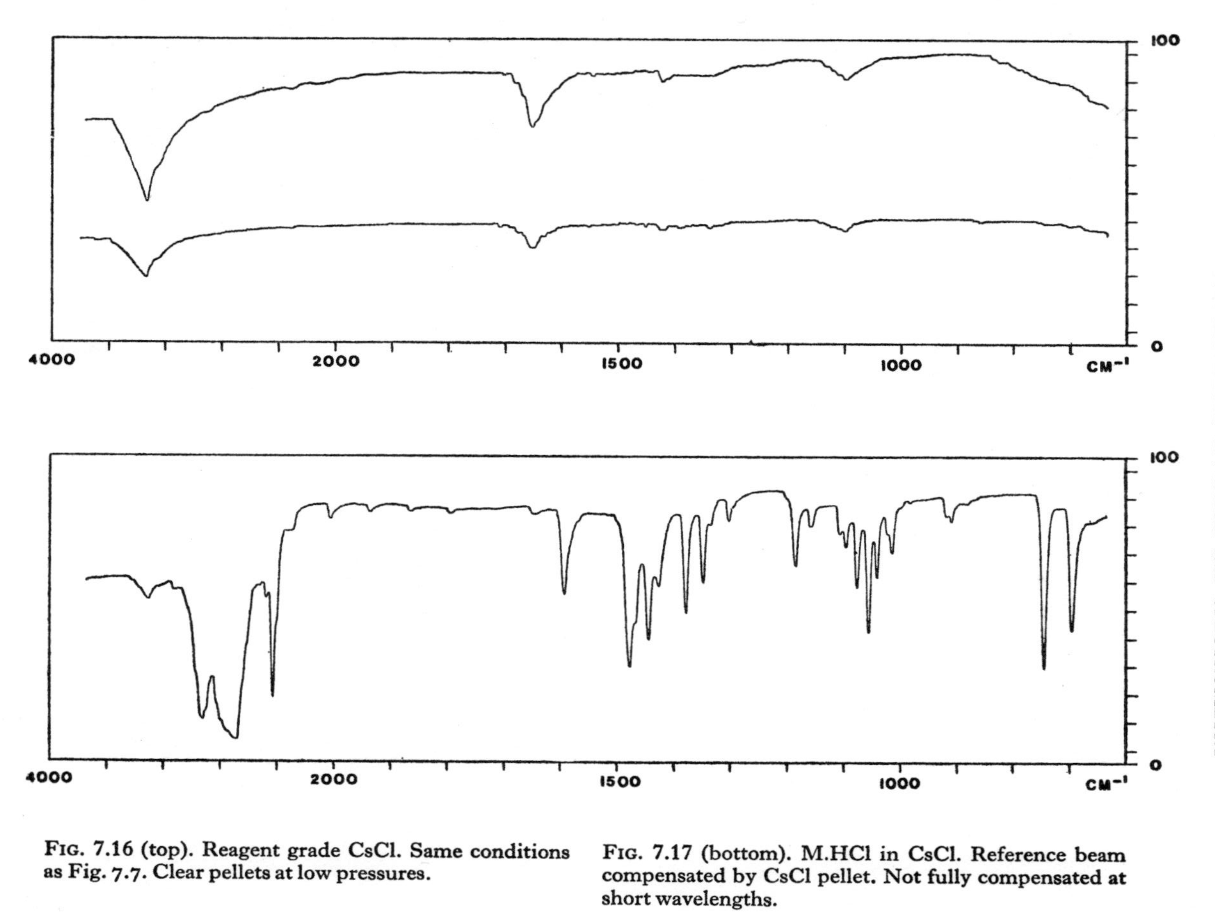

FIG. 7.16 (top). Reagent grade CsCl. Same conditions as Fig. 7.7. Clear pellets at low pressures.

FIG. 7.17 (bottom). M.HCl in CsCl. Reference beam compensated by CsCl pellet. Not fully compensated at short wavelengths.

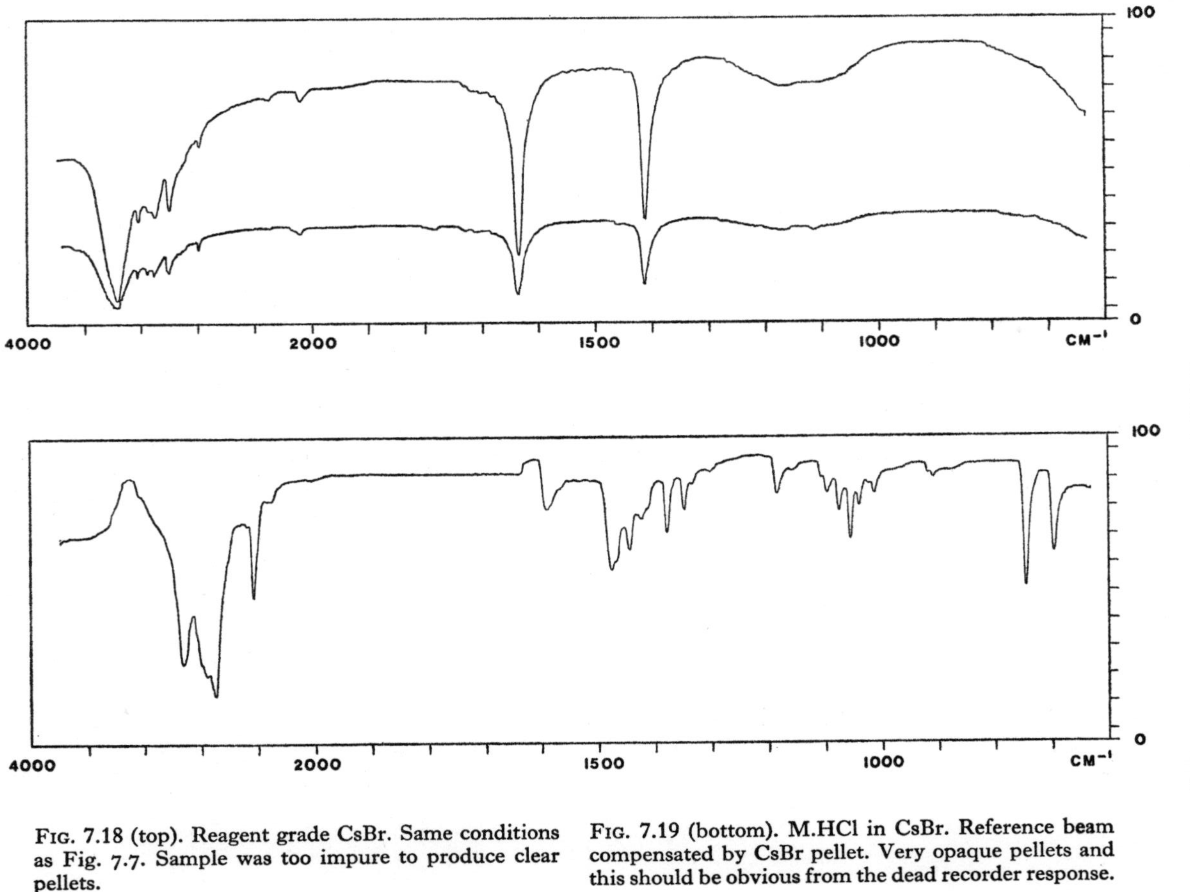

FIG. 7.18 (top). Reagent grade CsBr. Same conditions as Fig. 7.7. Sample was too impure to produce clear pellets.

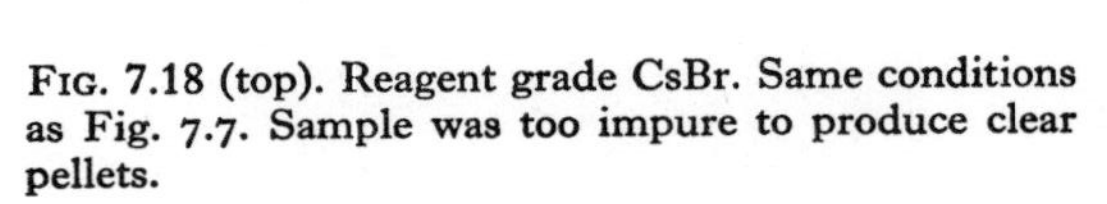

FIG. 7.19 (bottom). M.HCl in CsBr. Reference beam compensated by CsBr pellet. Very opaque pellets and this should be obvious from the dead recorder response.

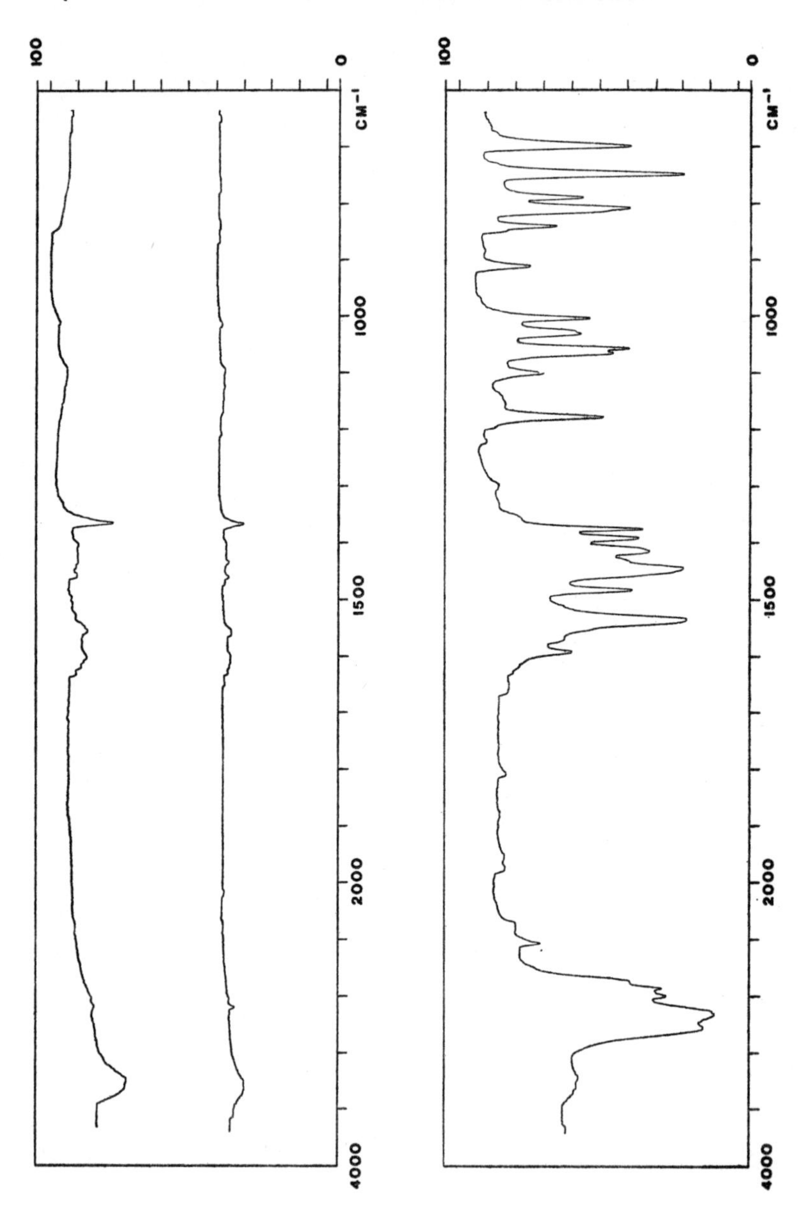
100
0
CM⁻¹
1000
1500
2000
4000
100
0
CM⁻¹
1000
1500
2000
4000

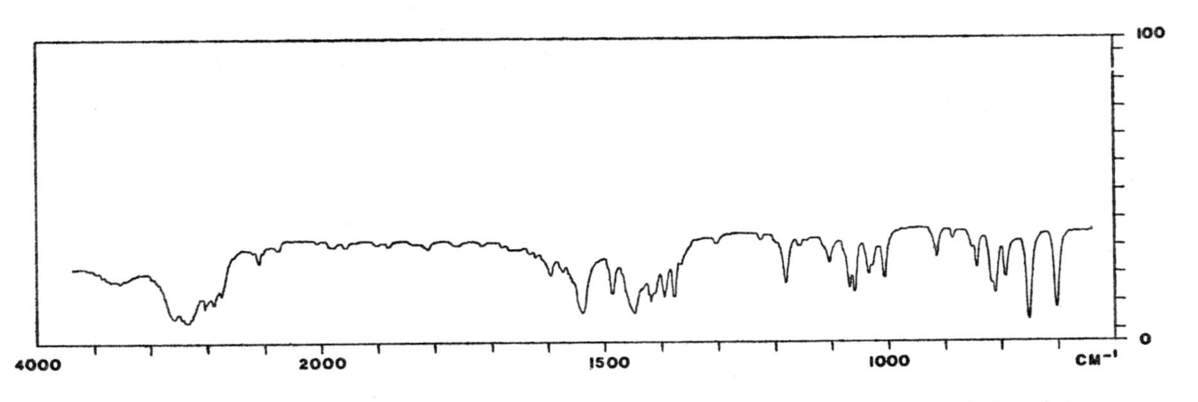

FIG. 7.20 (top). Analytical Reagent grade KI. Same conditions as Fig. 7.7. Sinters very readily to clear disks with a yellow tinge.

FIG. 7.21(*a*) (middle). M.HCl in KI. Reference beam compensated by KI pellet. The pellet has an even, yellow, opalescence with a high energy transmission.

FIG. 7.21(*b*) (bottom). As (*a*), except there is no reference beam compensation.

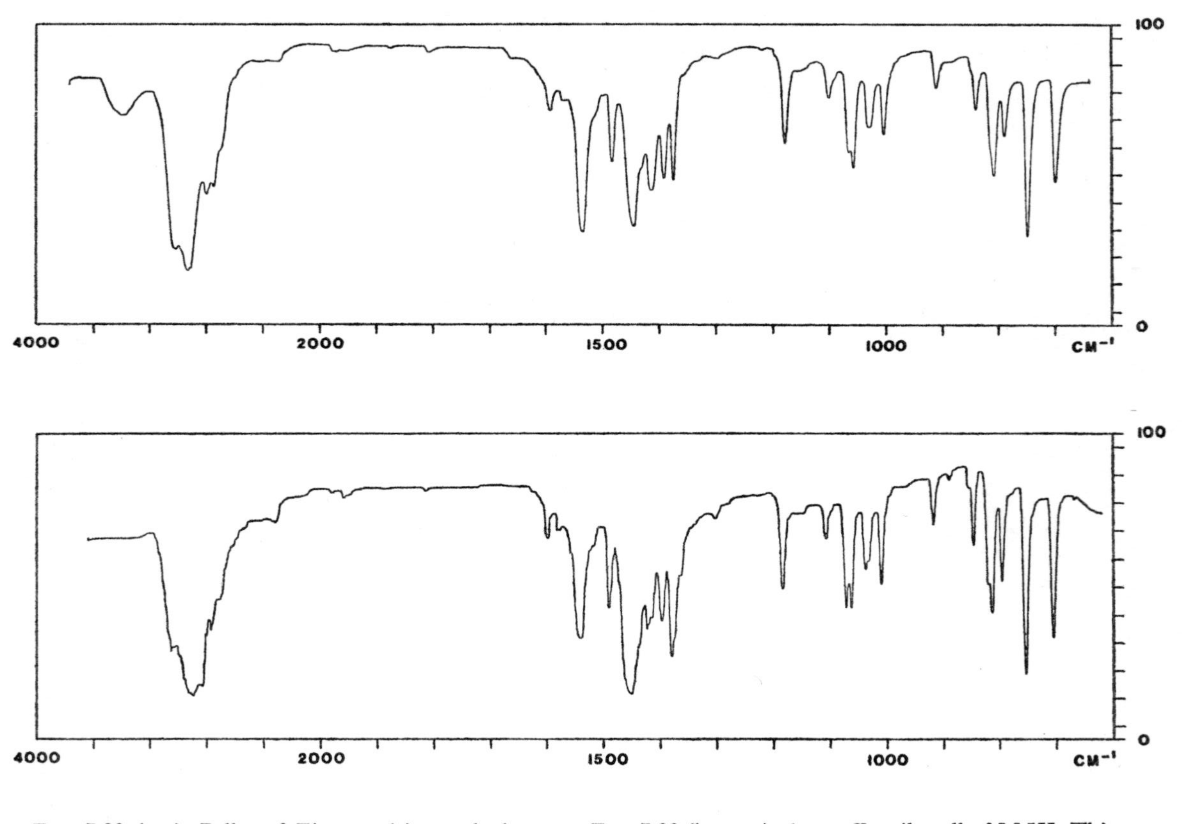

FIG. 7.22 (top). Pellet of Fig. 7.21(*a*) punched out, reground, and repressed. Some material lost in the process.

FIG. 7.23 (bottom). A paraffin oil mull of M.HI. This confirms that Figs. 7.21 and 7.22 show evidence for incomplete and complete ion-exchange respectively. M.HI is a very soft crystalline solid which is difficult to mull. Some light scatter is apparent.

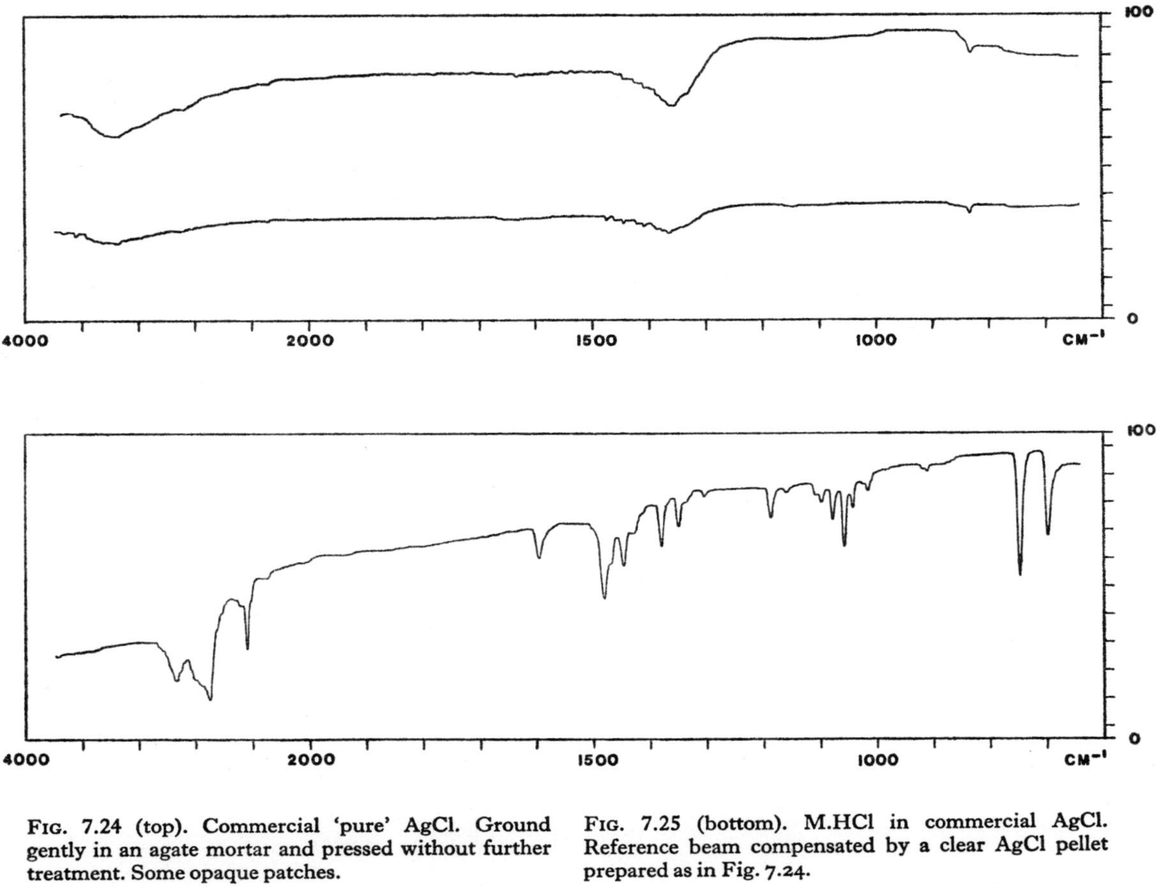

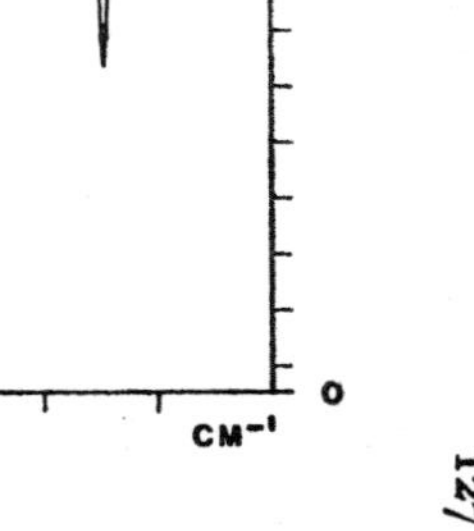

FIG. 7.24 (top). Commercial 'pure' AgCl. Ground gently in an agate mortar and pressed without further treatment. Some opaque patches.

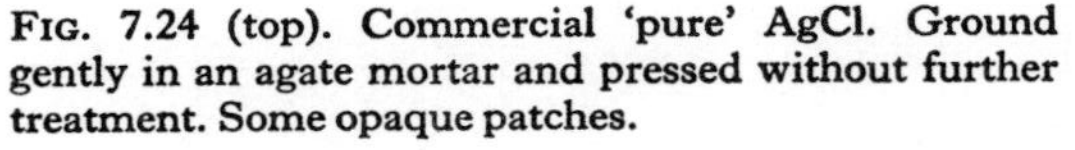

FIG. 7.25 (bottom). M.HCl in commercial AgCl. Reference beam compensated by a clear AgCl pellet prepared as in Fig. 7.24.

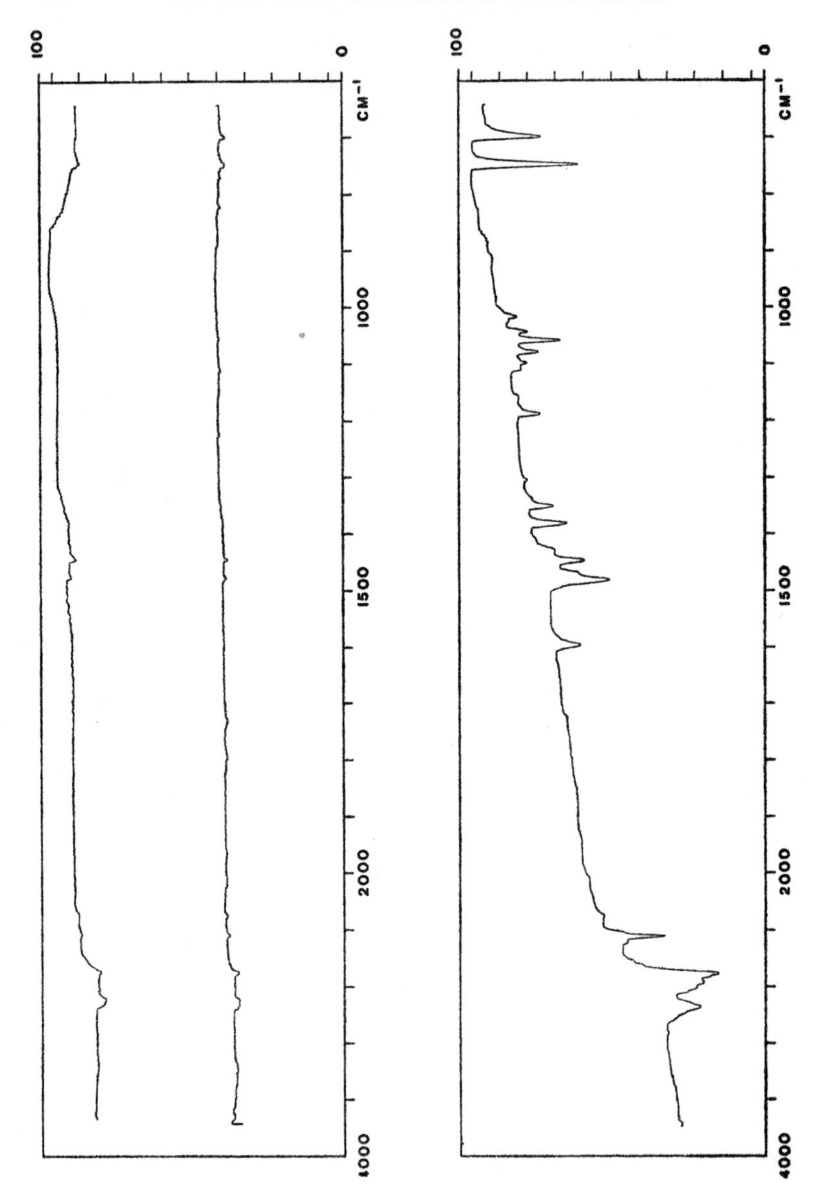
100
0
CM⁻¹
1000
1500
2000
4000
100
0
CM⁻¹
1000
1500
2000
4000

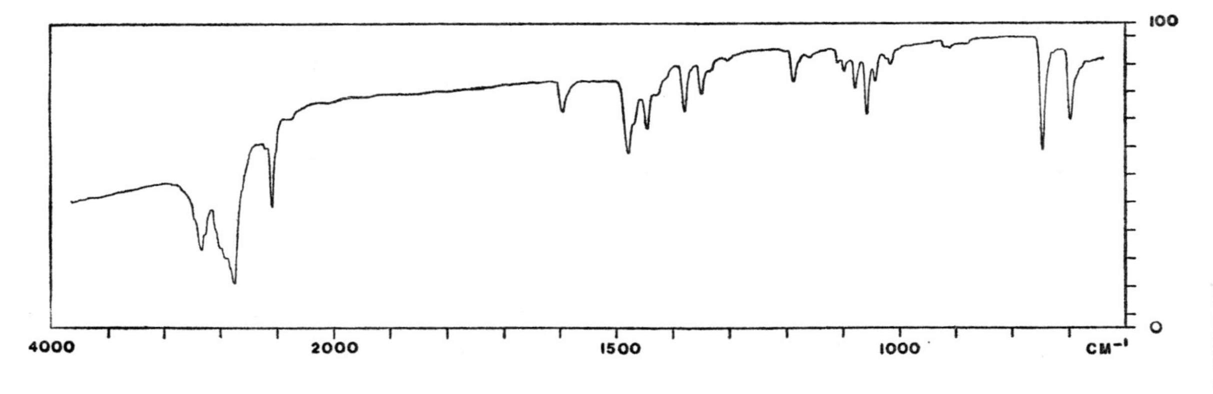

Fig. 7.26 (top). Precipitated AgCl, dried in the dark at 110°C. Ground gently in an agate mortar to a fine powder and pressed without further treatment.

Fig. 7.27 (middle). M.HCl in precipitated AgCl. Reference beam compensated by a clear pellet prepared as in Fig. 7.26. Pellet has an even opaque appearance.

Fig. 7.28 (bottom). Sample of Fig. 7.27 punched out, reground, and repressed with same compensation as Fig. 7.27. Pellet remains opaque, but transmission is noticeably better. Second grinding of matrix is very simple as this breaks down without clumping.

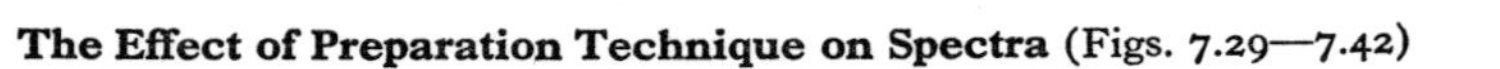

The Effect of Preparation Technique on Spectra (Figs. 7.29—7.42)

A batch of M.HCl/KCl was made up and evaporated from aqueous solution for spectra 7.29–7.33. The additional preparation is as noted.

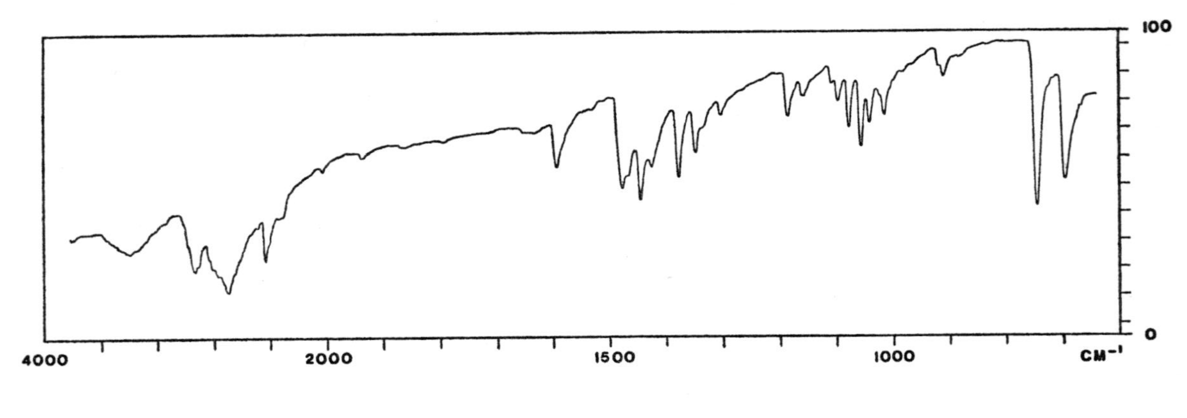

FIG. 7.29. Evaporated product pressed as received without any further preparation. An opaque pellet!

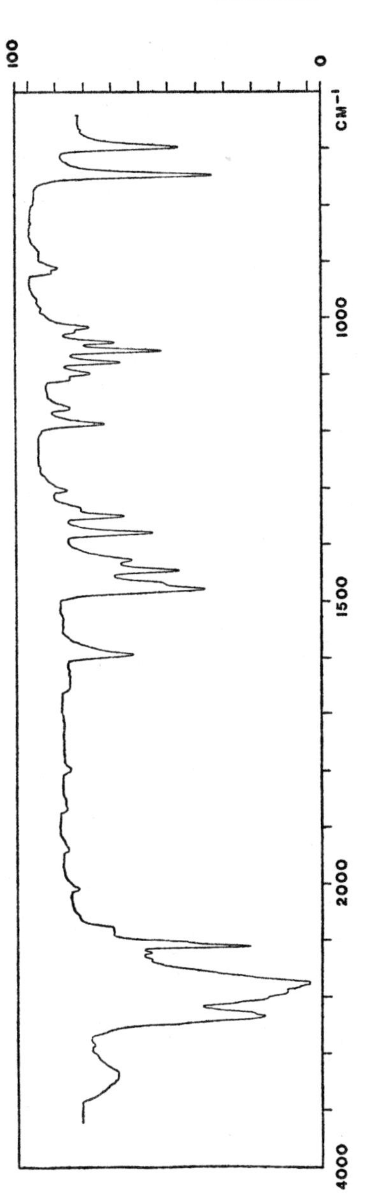

Fig. 7.30. Pellet of Fig. 7.29 punched out, reground very roughly to an even powder, and repressed. A clear, clean pellet, without excessive moisture content.

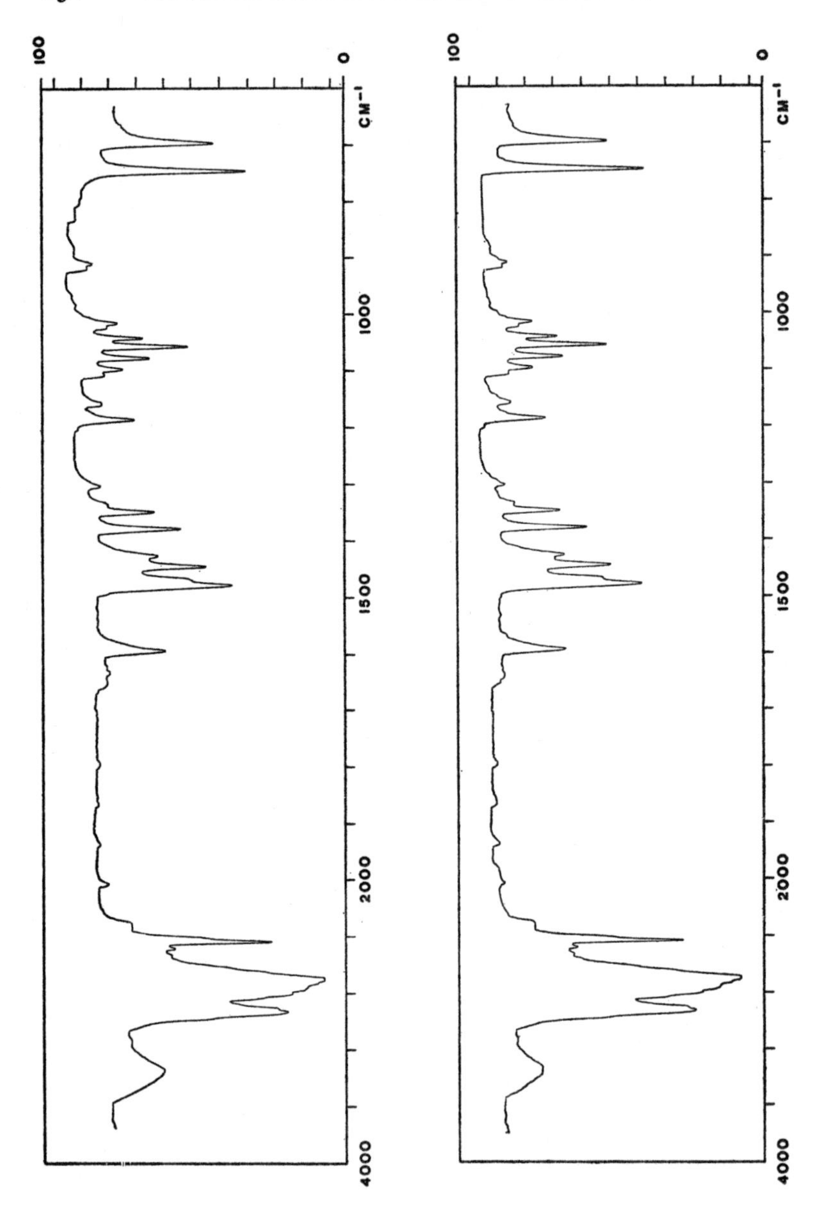
100
0
CM⁻¹
1000
1500
2000
4000
100
0
CM⁻¹
1000
1500
2000
4000

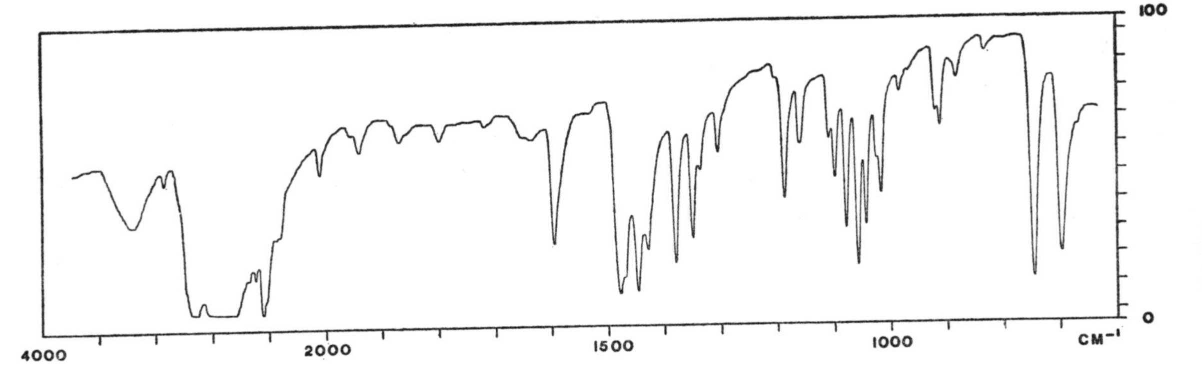

FIG. 7.31 (top). Pellet of Fig. 7.30 punched out, reground very roughly to an even powder, and repressed. Pellet remains clear, but the extra moisture content has had its effect, as well as the excessive grinding which has caused some fissures in the pellet with some energy scatter.

FIG. 7.32 (middle). Pellet of Fig. 7.31 fully compensated by a blank, slightly opaque pellet. Water is not fully compensated.

FIG. 7.33 (bottom). Evaporated product ground before pressing. A very thick pellet which is optically clear chosen for comparison. Relatively little energy scatter. No reference beam compensation.

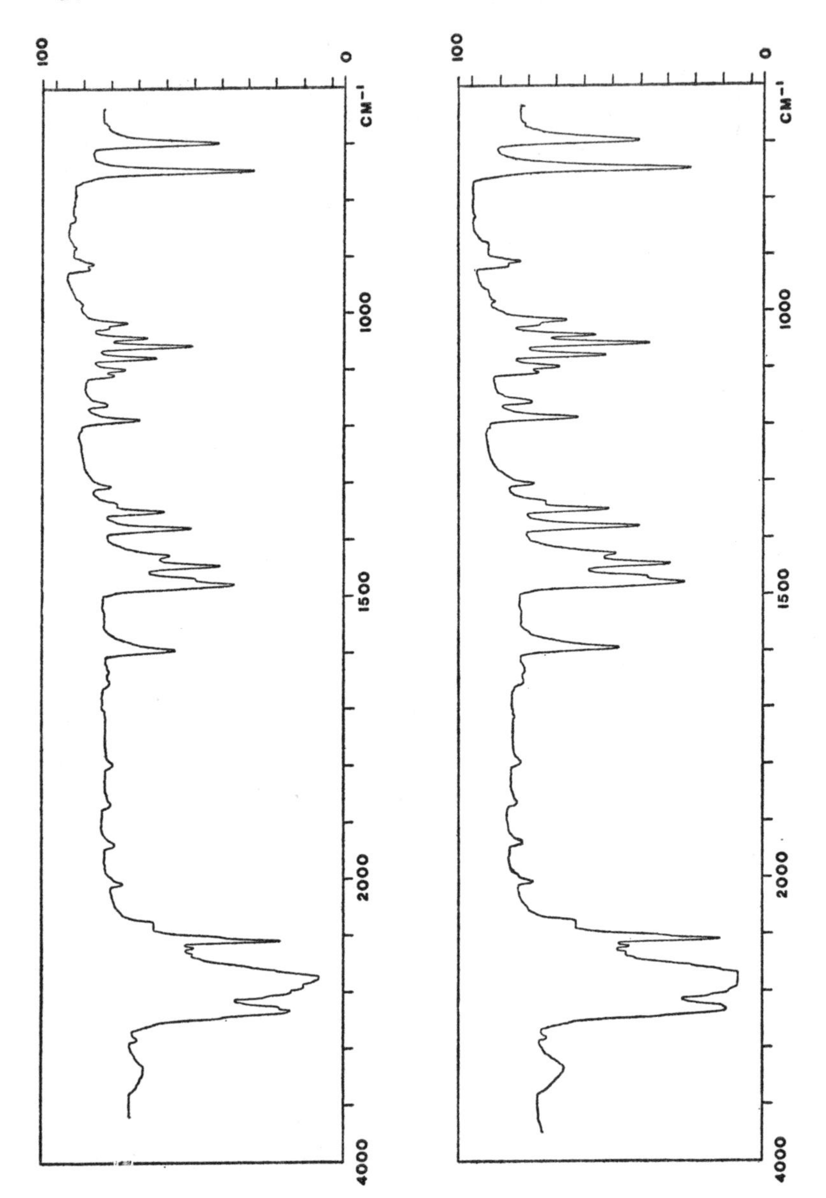
100
0
CM⁻¹
1000
1500
2000
4000
100
0
CM⁻¹
1000
1500
2000
4000

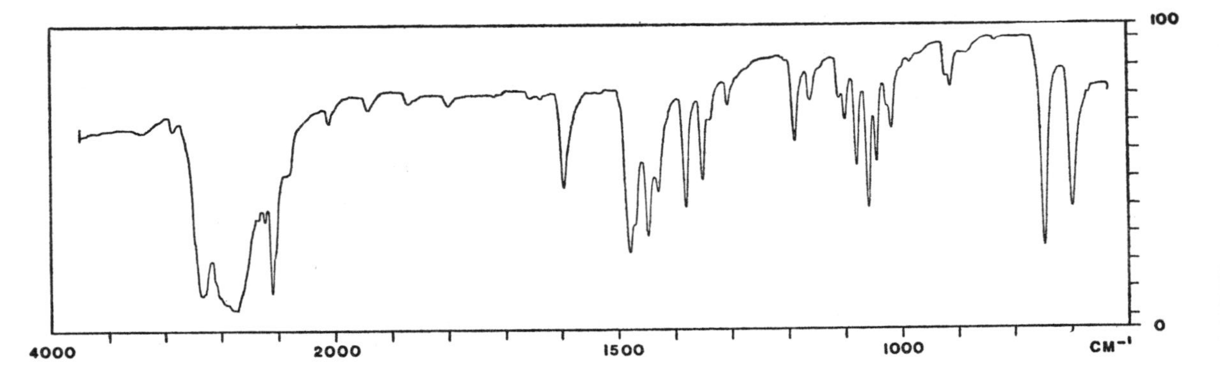

Fig. 7.34 (top). Solution of M.HCl in chloroform deposited on 100–204 mesh(BSI) powdered KCl. Product allowed to evaporate naturally and pressed without further treatment.

Fig. 7.35 (middle). Mixture prepared as in Fig. 7.34, except that it was ground before pressing. Apart from a slightly thicker pellet being used, there is no obvious advantage in grinding after deposition; the moisture content is increased.

Fig. 7.36 (bottom). Solution of M.HCl in chloroform deposited on powdered KCl finer than 240 mesh(BSI). Product allowed to evaporate naturally and pressed without further treatment. Same quantities as in Fig. 7.35.

The spectrum is not improved and fissures are produced in the pellet which is obvious from the higher energy dispersion at short wavelengths.

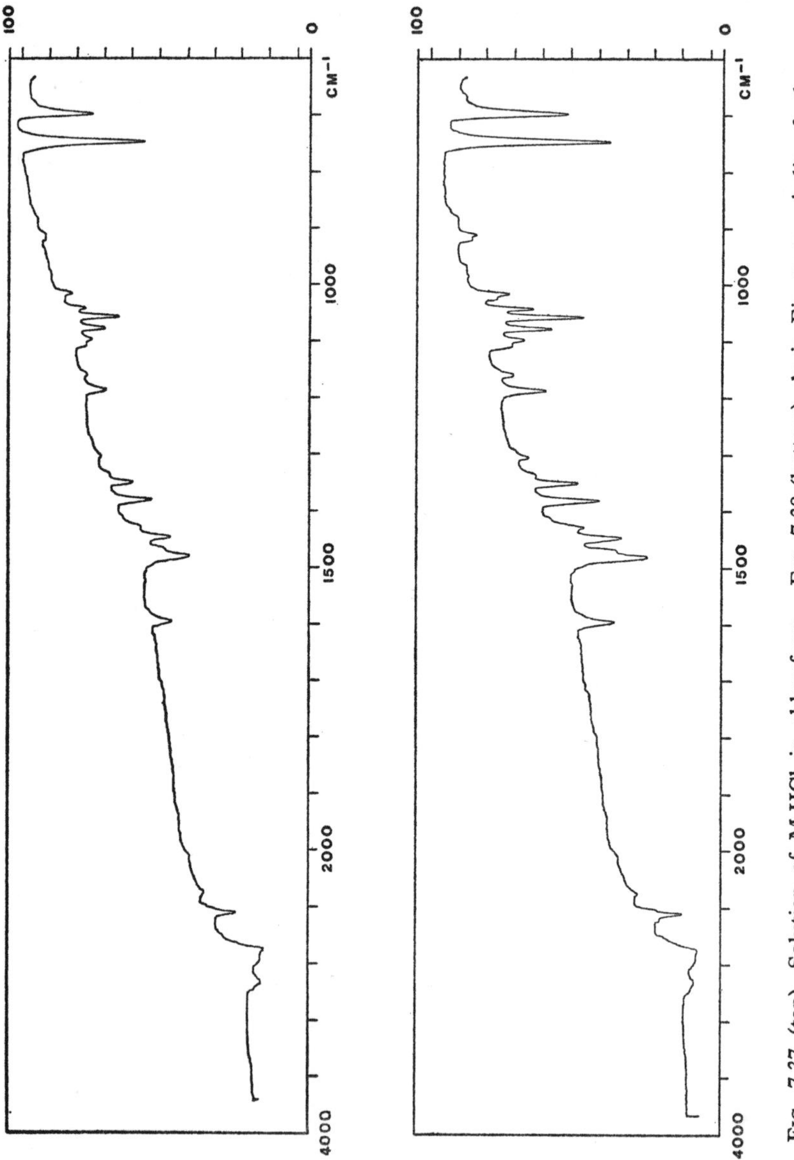

FIG. 7.37 (top). Solution of M.HCl in chloroform deposited on finely powdered, precipitated silver chloride. After evaporation, pellet pressed without further treatment.

FIG. 7.38 (bottom). As in Fig. 7.37, grinding further before pressing. Although a higher concentration than Fig. 7.37, there is no marked improvement. Figs. 7.37 and 7.38 are better in some respects than Fig. 7.28, even though energy scatter is more pronounced here.

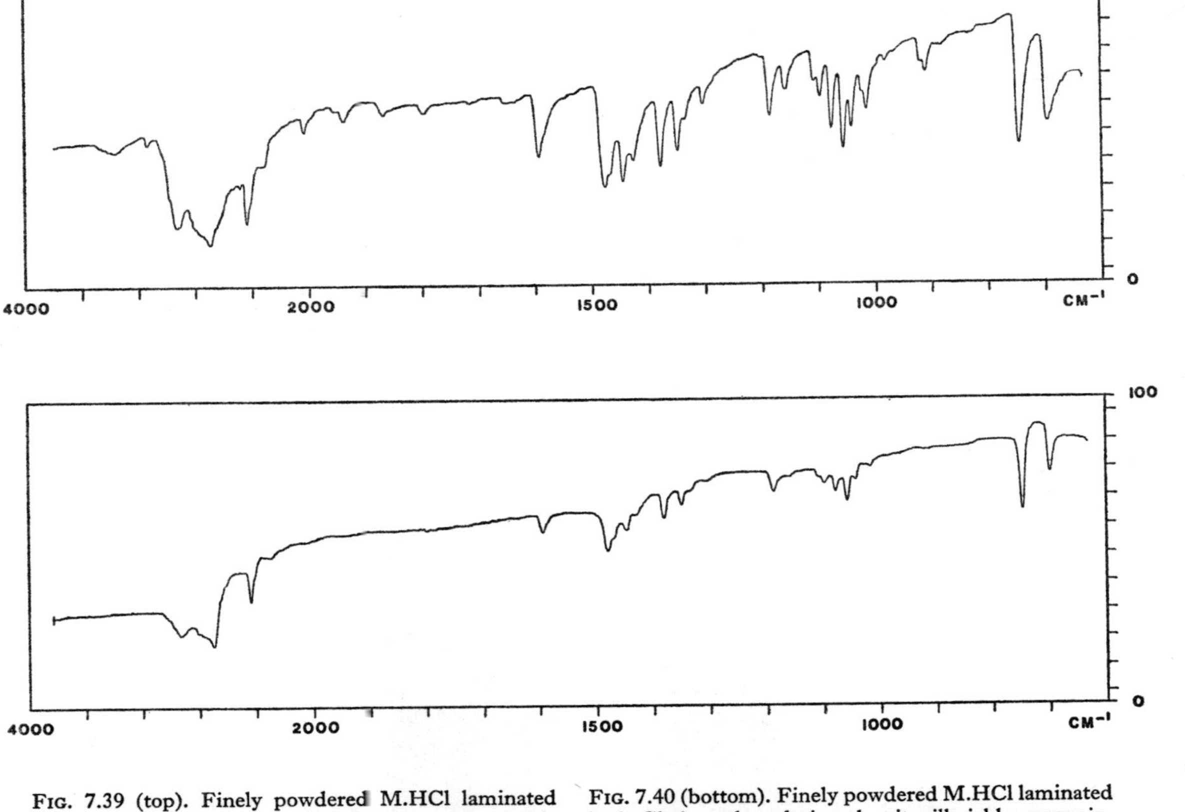

FIG. 7.39 (top). Finely powdered M.HCl laminated in KCl. Large energy scatter but a recognizable spectrum if base-line drift is allowed for.

FIG. 7.40 (bottom). Finely powdered M.HCl laminated in AgCl. A crude technique but it will yield a recognizable spectrum, even at the low levels used here.

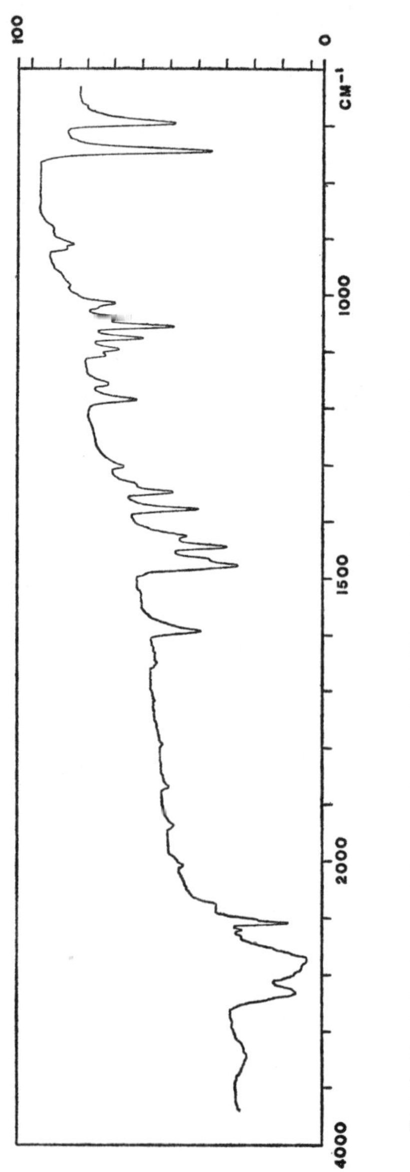

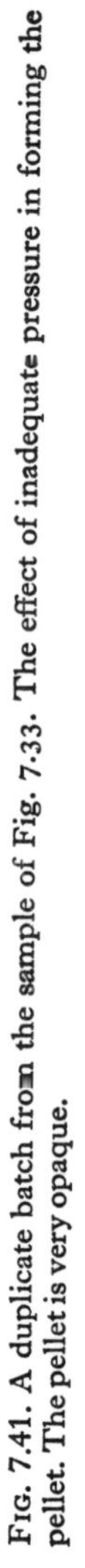

FIG. 7.41. A duplicate batch from the sample of Fig. 7·33. The effect of inadequate pressure in forming the pellet. The pellet is very opaque.

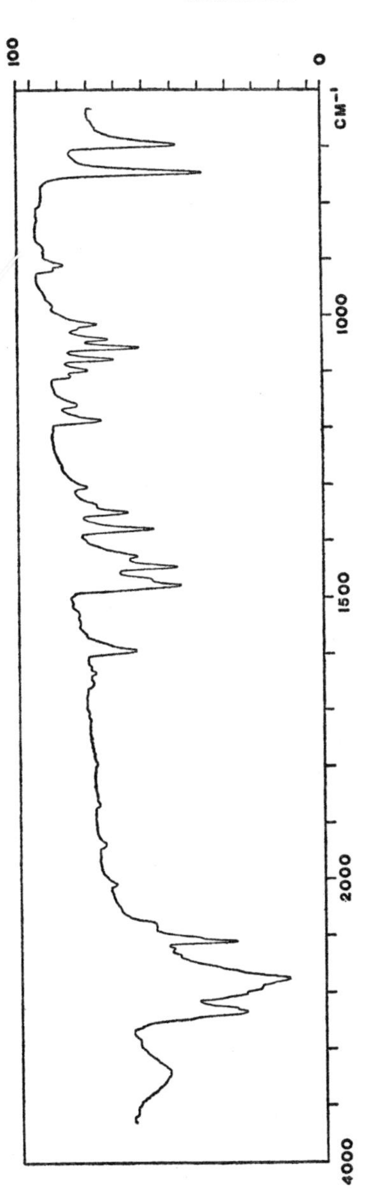

FIG. 7.42. A duplicate batch from Fig. 7.33. The effect of moisture caused by leaving the ground product in the atmosphere for an excessive period. There are isolated opaque spots in this pellet.

Paraffin Oil Mulls (Figs. 7.43—7.45)

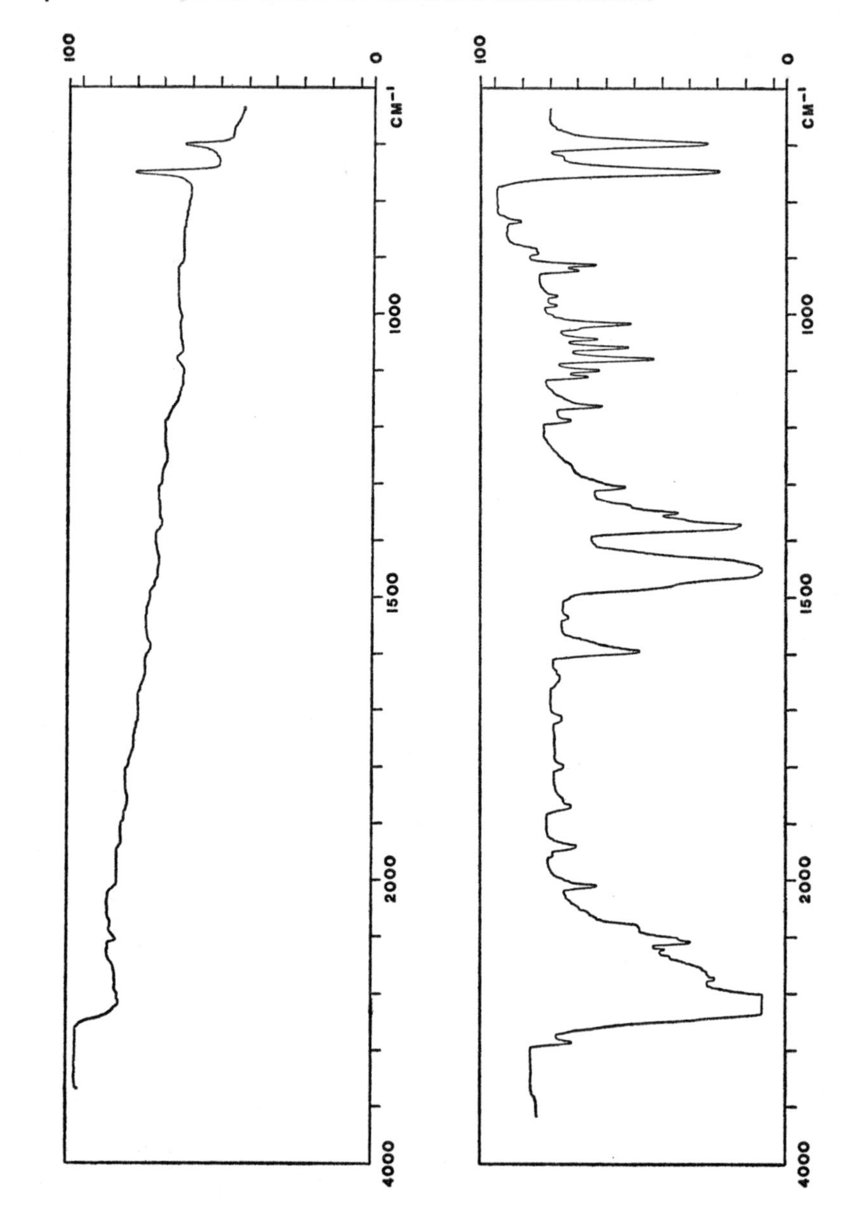

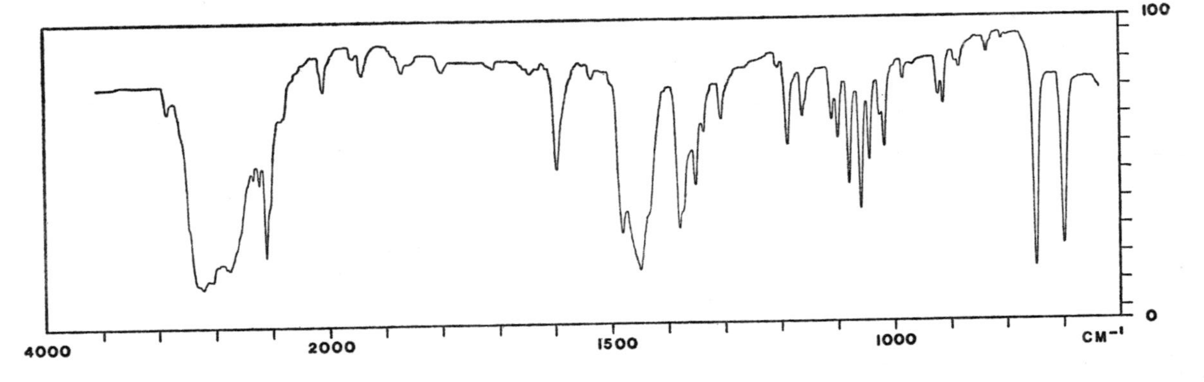

FIG. 7.43 (top). Finely powdered M.HCl deposited on an NaCl plate.

FIG. 7.44 (middle). Crystals of M.HCl from the reagent bottle, moistened with paraffin oil and retained between two NaCl plates. This spectrum is better than some of the pressed-disk spectra, even though the material has not been ground. The bands due to the crystal structure are obvious, and should be used to criticize technique in the other spectra.

FIG. 7.45 (bottom). The mixture of Fig. 7.44 ground slightly. This approximates to the normal mulling technique, where grinding takes place under oil.

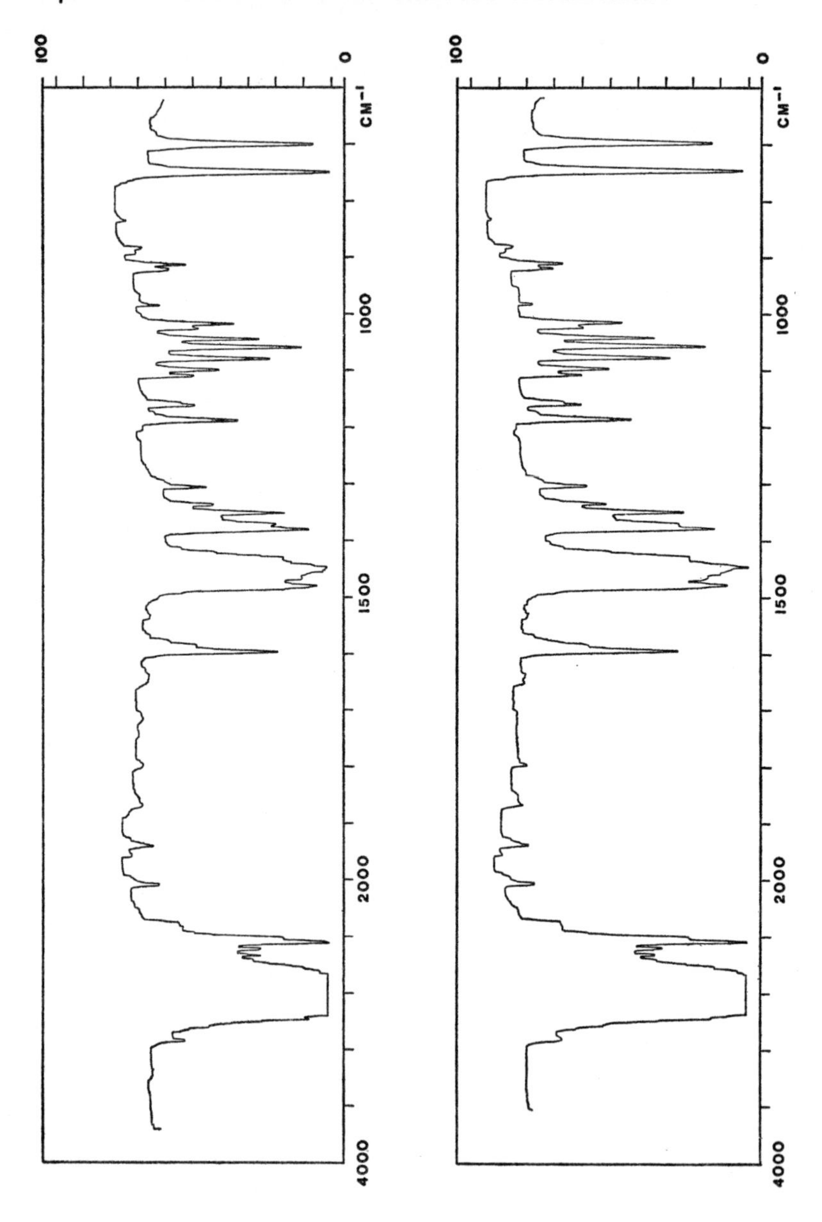
100
0
CM⁻¹
1000
1500
2000
4000
100
0
CM⁻¹
1000
1500
2000
4000

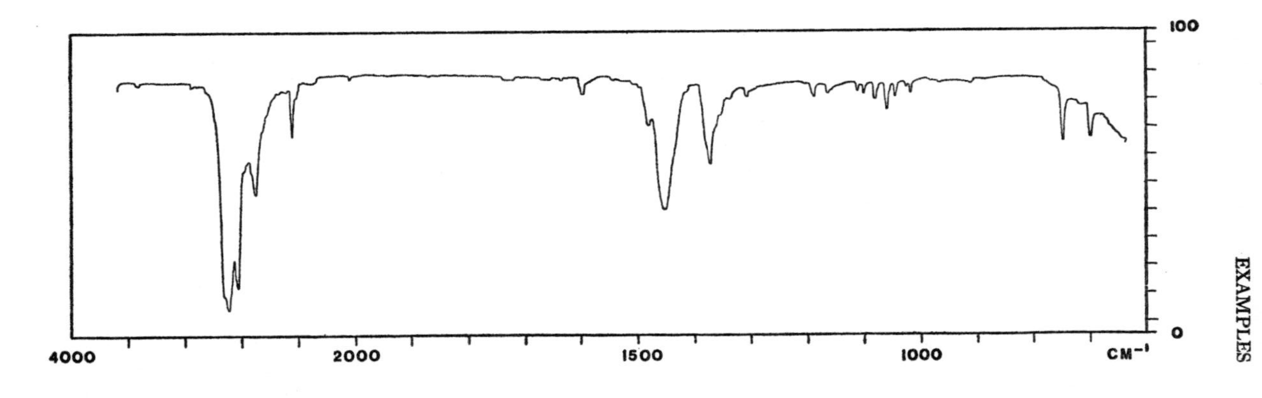

FIG. 7.46 (top). A reasonable mull. Prepared by prior grinding. No compensation in the reference beam.

FIG. 7.47 (middle). The mull of Fig. 7.44 with attenuated reference beam.

FIG. 7.48 (bottom). The mull of Fig. 7.44, thinned with further oil. No reference beam compensation.

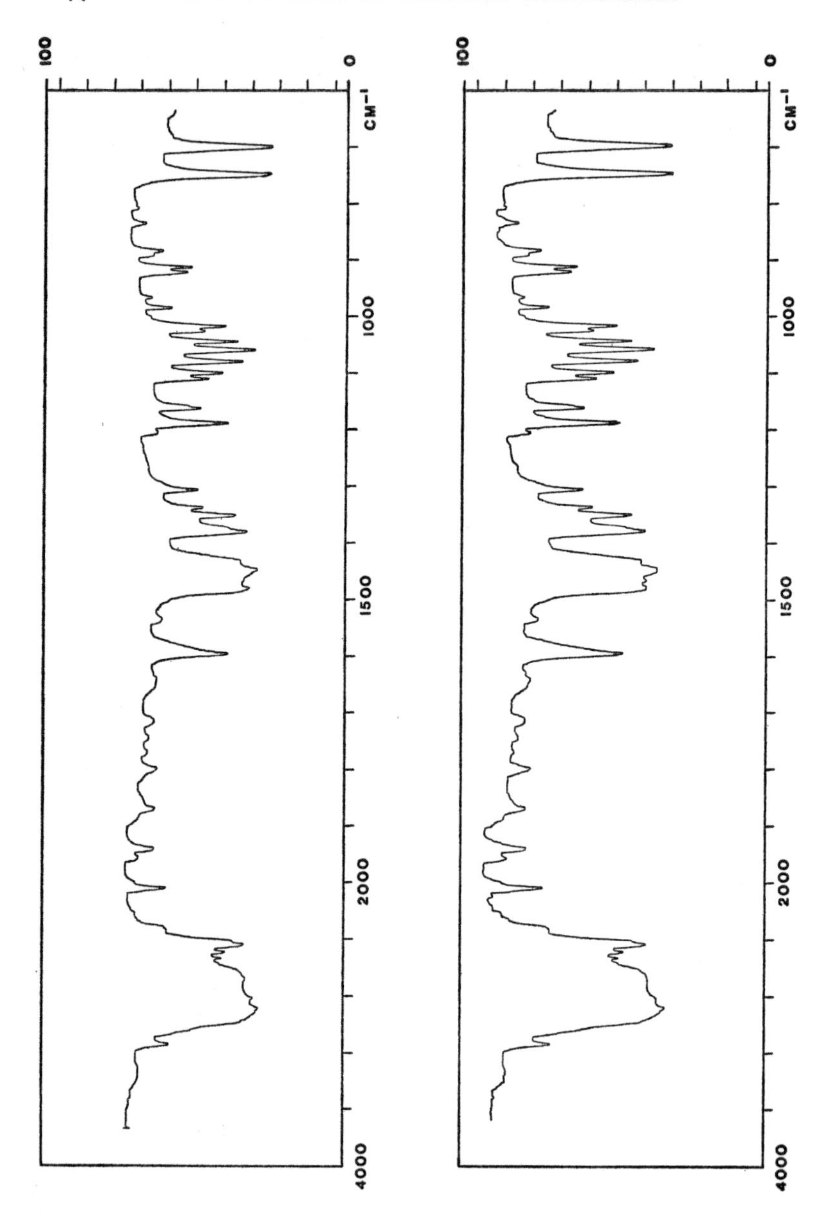
100
0
CM⁻¹
1000
1500
2000
4000
100
0
CM⁻¹
1000
1500
2000
4000

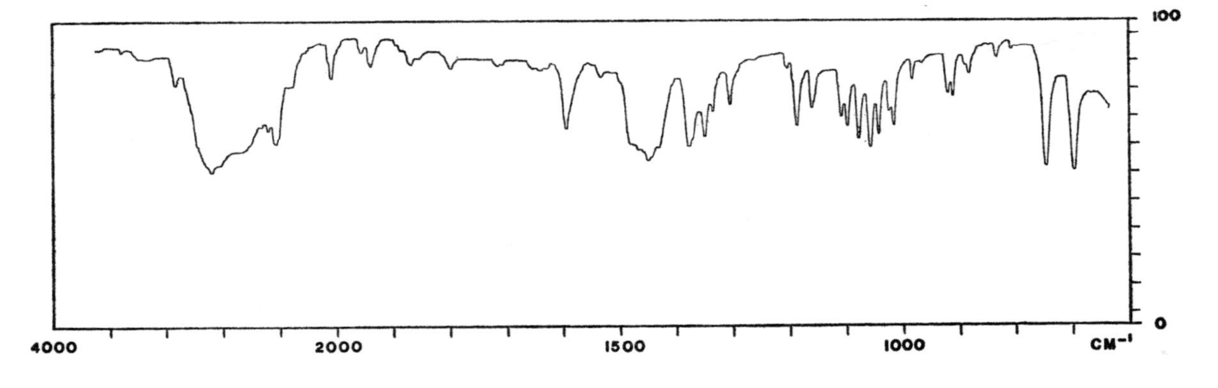

FIG. 7.49 (top). A stiff mull, corresponding to the preparation of Fig. 7.45, incompletely filling the sample beam. Presented 'as is' without compensation.

FIG. 7.50. (middle). Mull of Fig. 7.49 with attenuated reference beam.

FIG. 7.51 (bottom). A more extreme example of Fig 7.50.

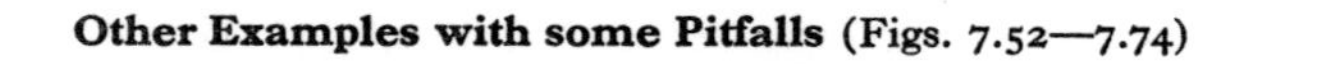

Other Examples with some Pitfalls (Figs. 7.52—7.74)

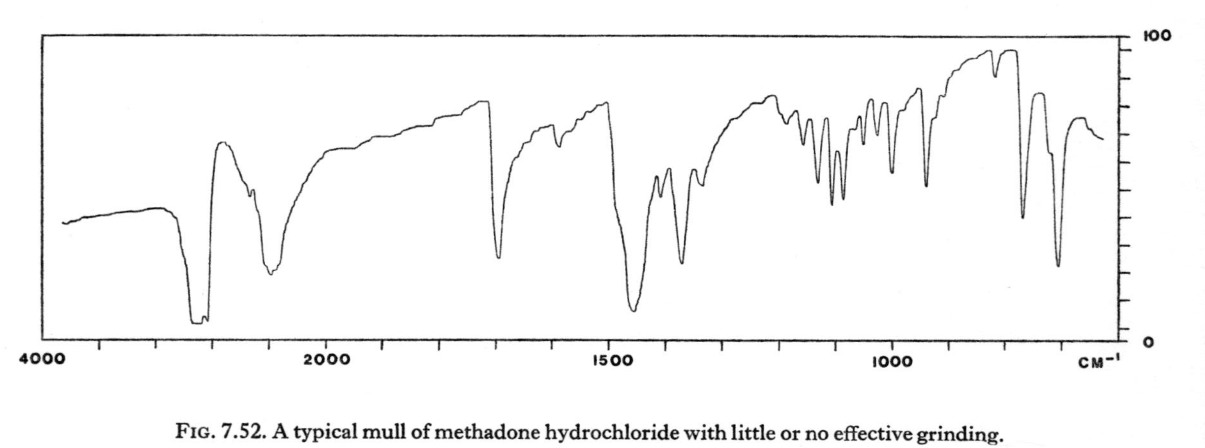

FIG. 7.52. A typical mull of methadone hydrochloride with little or no effective grinding.

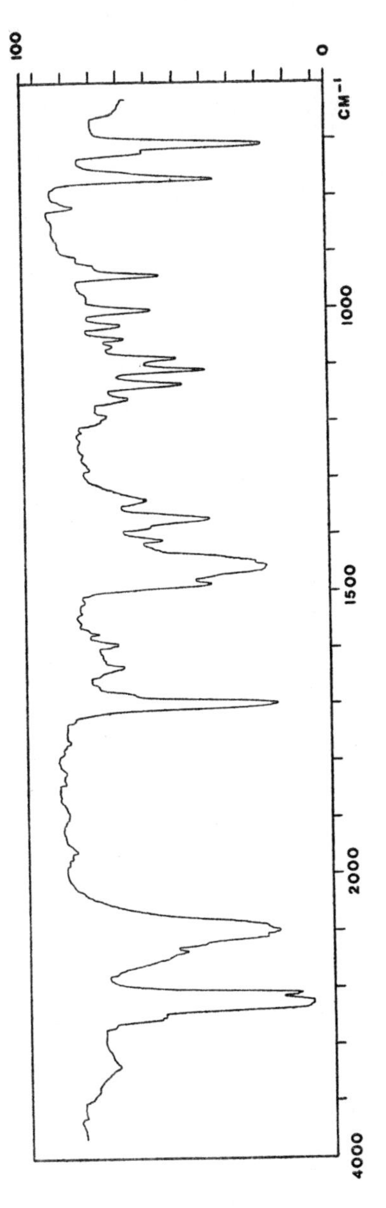

FIG. 7.53. A better mull. The bands in Fig. 7.52 appearing to suffer from Christiansen filter effects are not subject to any marked change or shift in frequency.

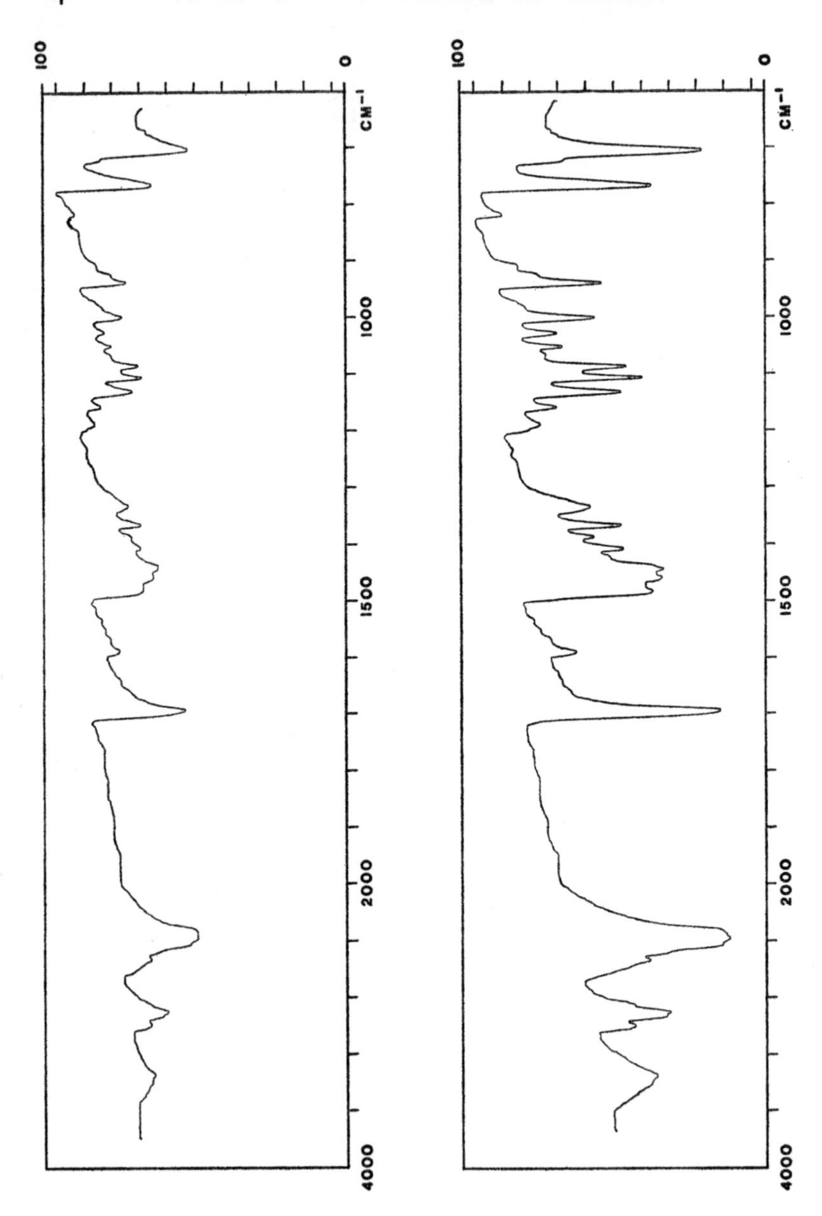
100
0
CM⁻¹
1000
1500
2000
4000
100
0
CM⁻¹
1000
1500
2000
4000

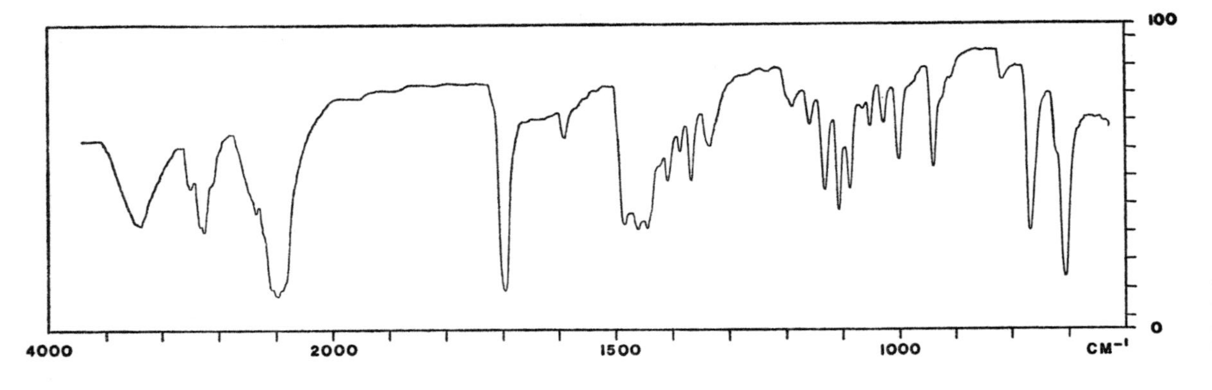

FIG. 7.54 (top). Methadone hydrochloride in KCl, with little or no grinding. The pellet is very clear, visually, with one or two isolated opaque spots. The true situation is revealed by the drifting baseline and the 'cramped' nature of the spectrum.

FIG. 7.55 (middle). The same pellet as Fig. 7.54, punched out, reground and repressed. The pellet is now translucent, indicating some incompatibility with the matrix; the Christiansen filter effect is not obvious in a marked change of frequency.

FIG. 7.56 (bottom). The pellet of Fig. 7.55 punched out, reground and repressed. The pellet remains translucent.

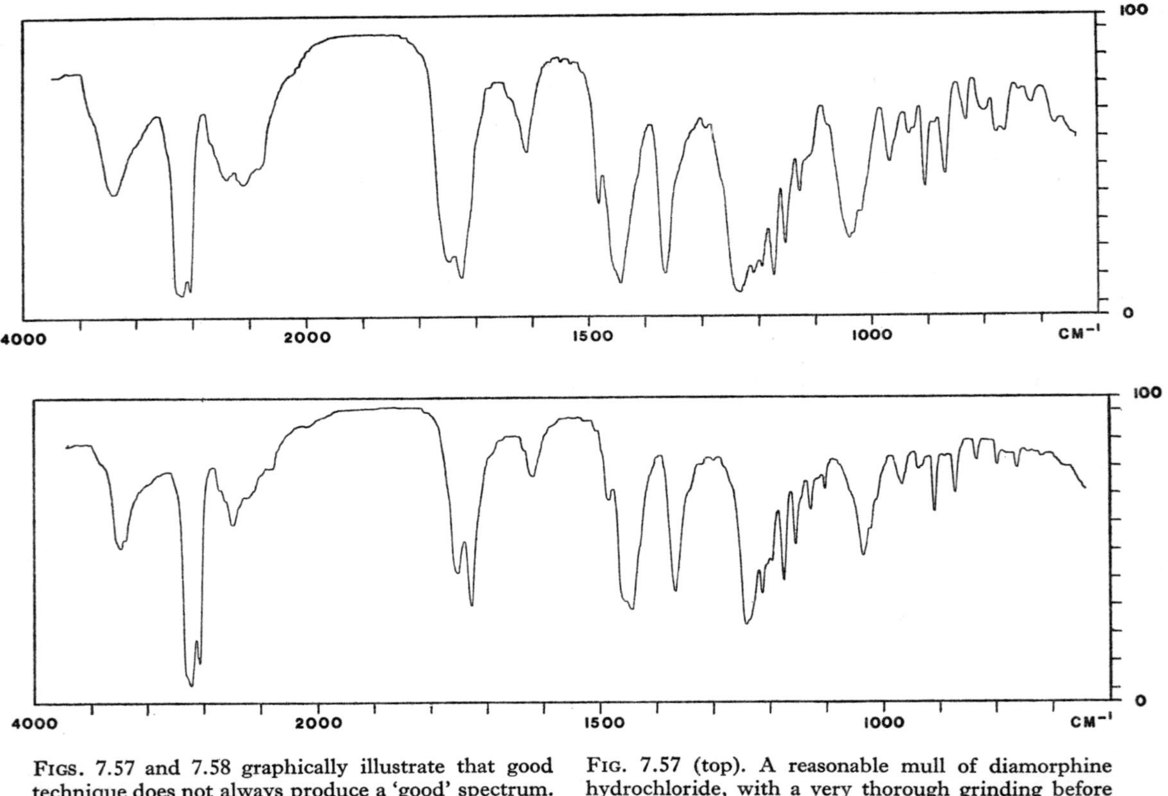

FIGS. 7.57 and 7.58 graphically illustrate that good technique does not always produce a 'good' spectrum. A fully ground powder is obviously converted to an amorphous polymorph, which is relatively stable over a period of time.

This is confirmed by leaving both mulls overnight.

The reasonable mull does not change much, whereas the truely poor mull appears to 'improve'.

FIG. 7.57 (top). A reasonable mull of diamorphine hydrochloride, with a very thorough grinding before mulling.

FIG. 7.58 (bottom). A mull of diamorphine hydrochloride prepared by the normal process, either by a perfunctory grinding before mulling, or by grinding under oil.

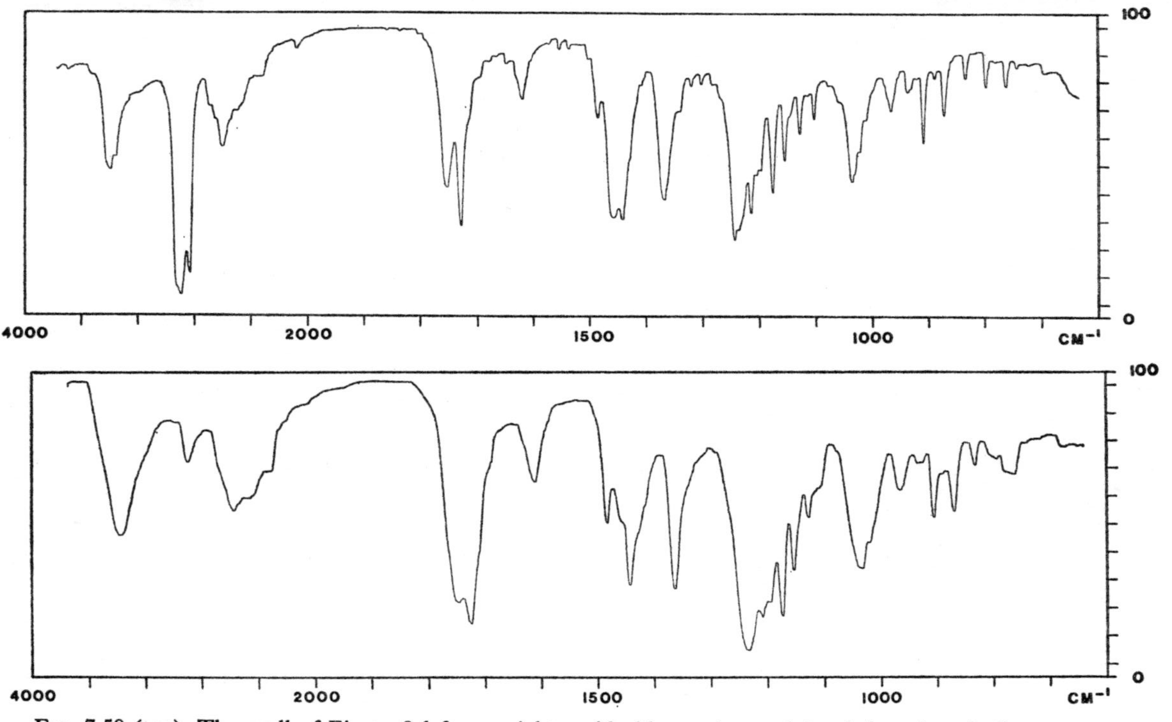

FIG. 7.59 (top). The mull of Fig. 7.58 left overnight. Amorphous material produced by partial grinding reverts to a crystalline polymorph.

Pressed disks are not alone in supplying examples of crystalline or polymorphic transitions.

Fully ground diamorphine hydrochloride is a fluffy powder and not a cake. Its colloidal nature is such that a mull is difficult to produce and the particles tend to stick together and will not disperse in the oil without considerable perseverance. Other alkaloid hydrochlorides, such as codeine, behave in a similar manner.

Diamorphine hydrochloride has been chosen to demonstrate the penalties of good technique and the importance of carrying out control experiments oneself because of the legal status of this drug. The effect on a retrieval system should be self evident.

FIG. 7.60 (bottom). A typical spectrum of diamorphine hydrochloride in KBr. Present as the amorphous form, which may not be recovered from a retrieval system based on mulls.

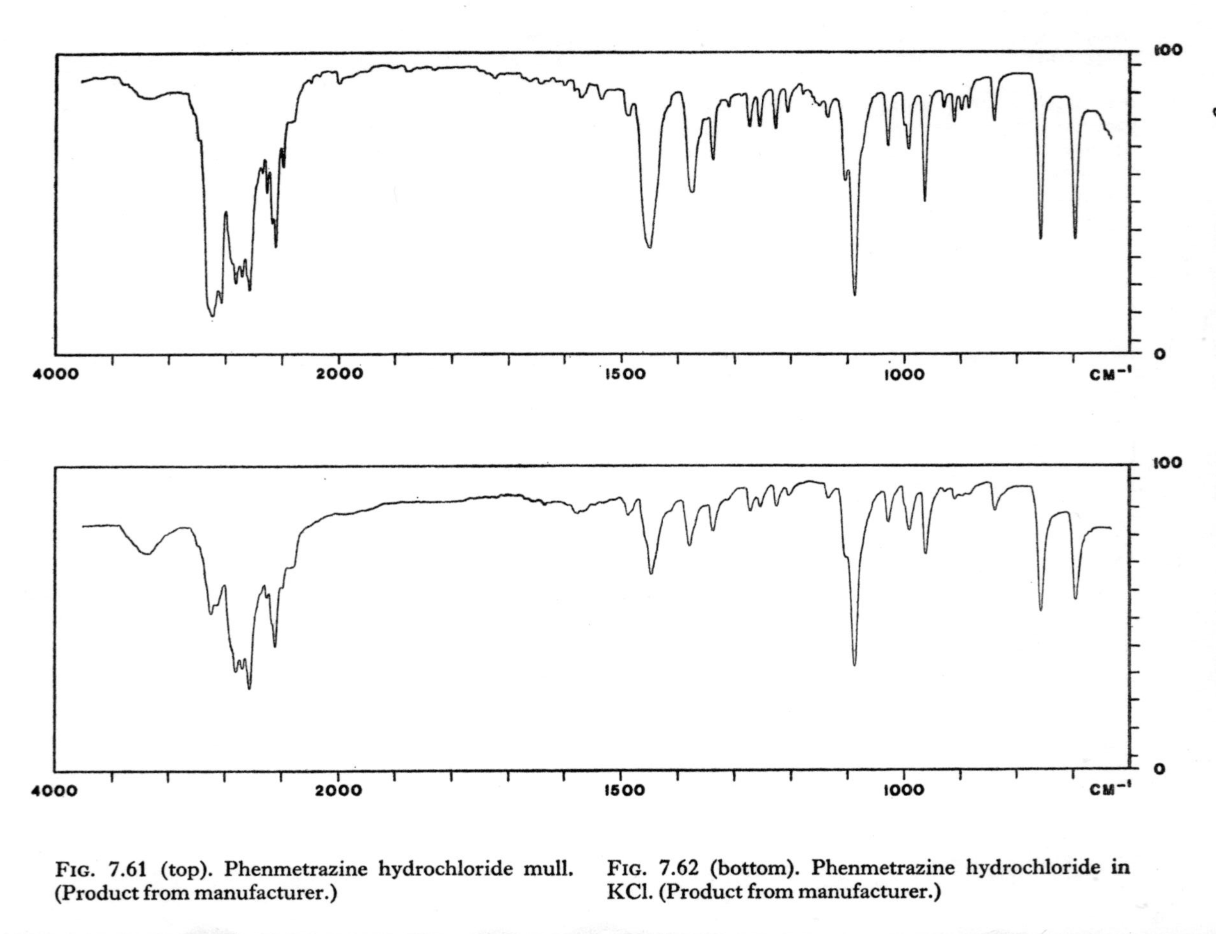

FIG. 7.61 (top). Phenmetrazine hydrochloride mull. (Product from manufacturer.)

FIG. 7.62 (bottom). Phenmetrazine hydrochloride in KCl. (Product from manufacturer.)

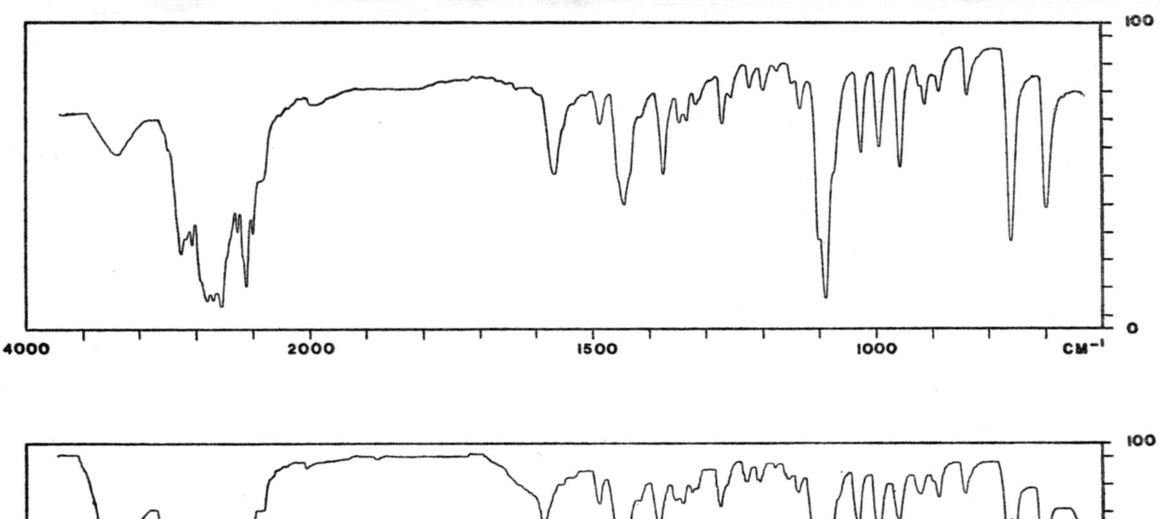

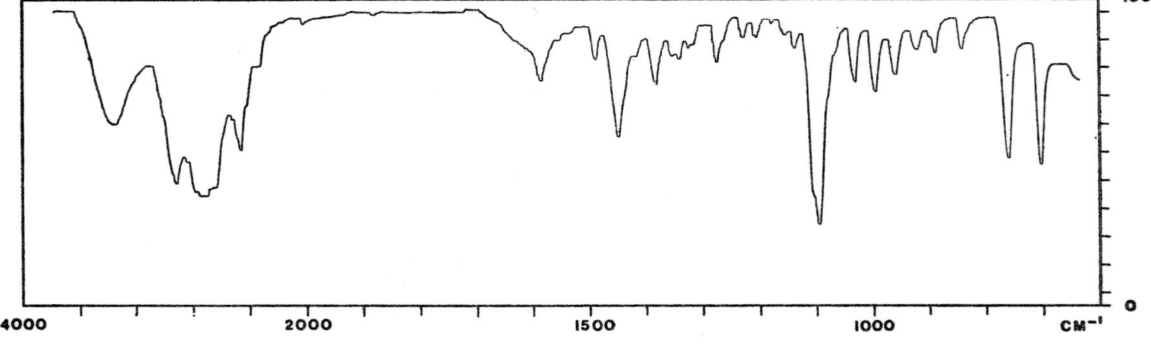

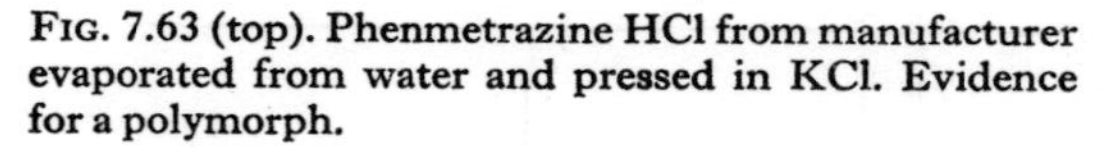
FIG. 7.63 (top). Phenmetrazine HCl from manufacturer evaporated from water and pressed in KCl. Evidence for a polymorph.

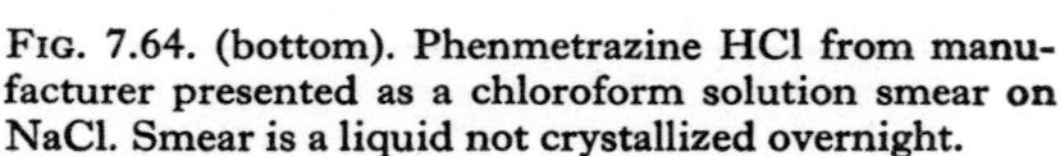
FIG. 7.64. (bottom). Phenmetrazine HCl from manufacturer presented as a chloroform solution smear on NaCl. Smear is a liquid not crystallized overnight.

F

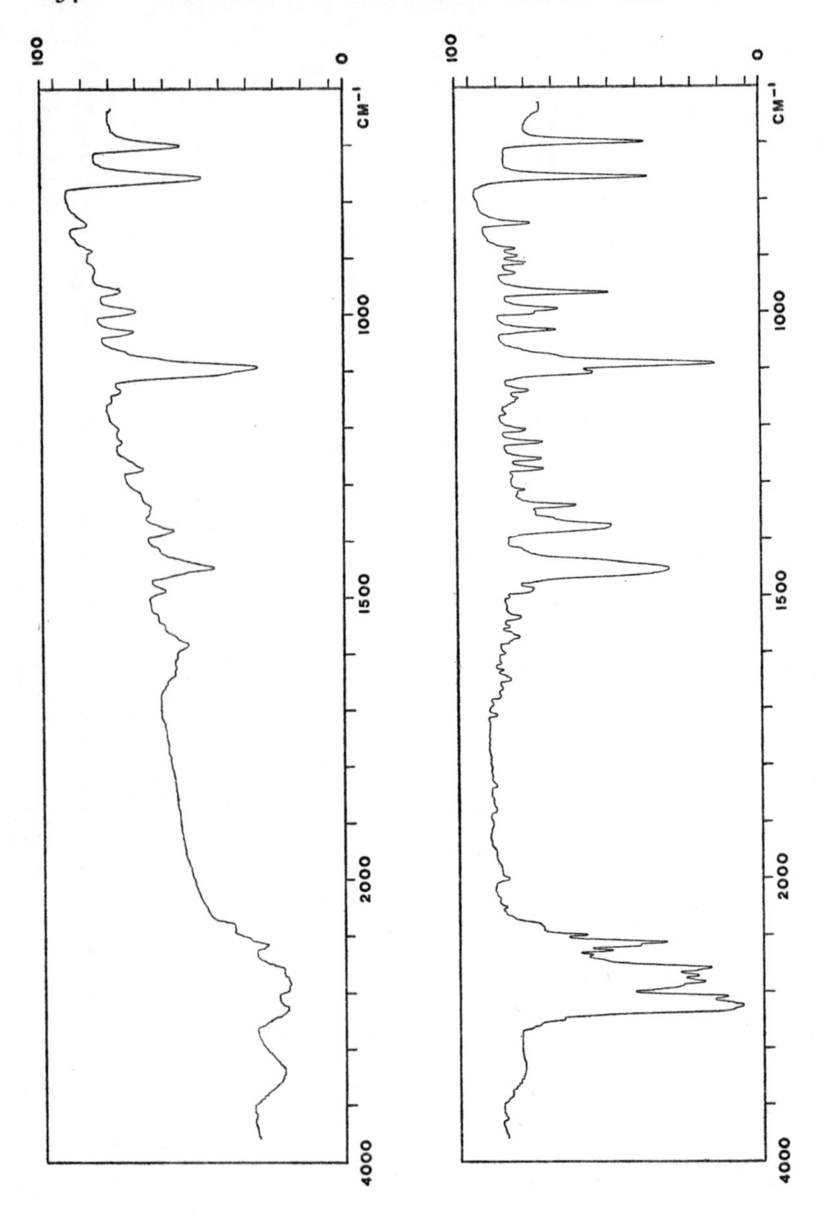
100
0
CM^{-1}
1000
1500
2000
4000
100
0
CM^{-1}
1000
1500
2000
4000

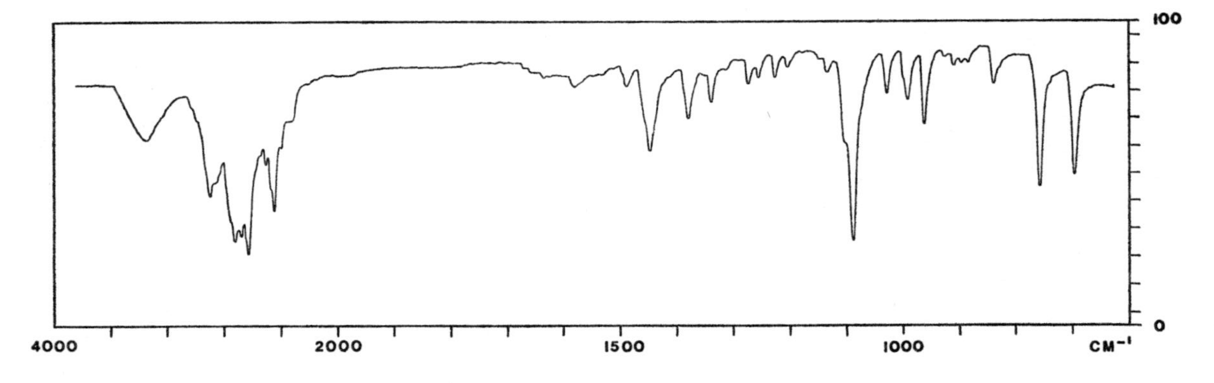

FIG. 7.65 (top). Phenmetrazine HCl solution in chloroform deposited on 100–240 mesh(BSI) KCl powder and pressed without further preparation. Some considerable opacity to the pellet and the material is present as the liquid polymorph. The pellet shows little change overnight.

This shows the importance of carrying out controls at all stages.

FIG. 7.66 (middle). Liquid phenmetrazine HCl polymorph (from chloroform solution) allowed to crystallize. This sample took over 1 week to crystallize. Presented as a mull.

FIG. 7.67 (bottom). The sample of Fig. 7.66 presented as a KCl disk.

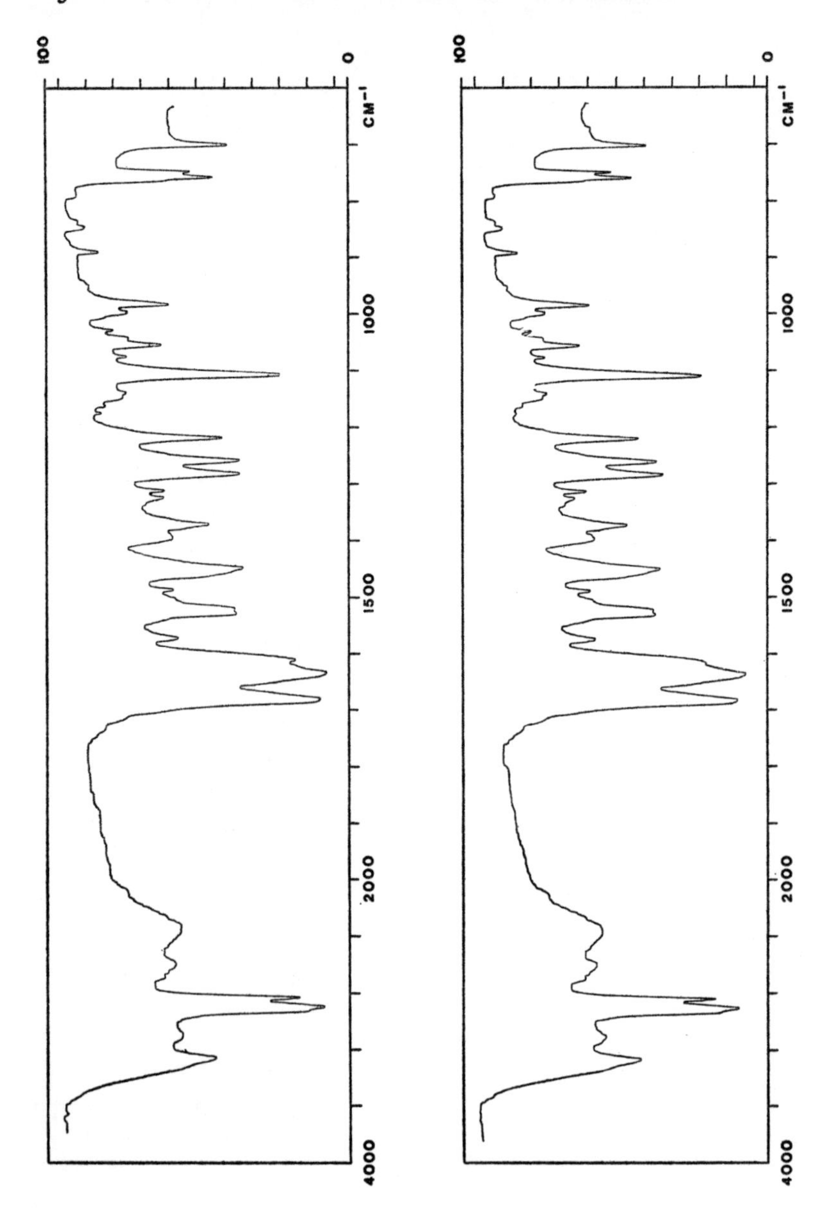
100
0
CM⁻¹
1000
1500
2000
4000
100
0
CM⁻¹
1000
1500
2000
4000

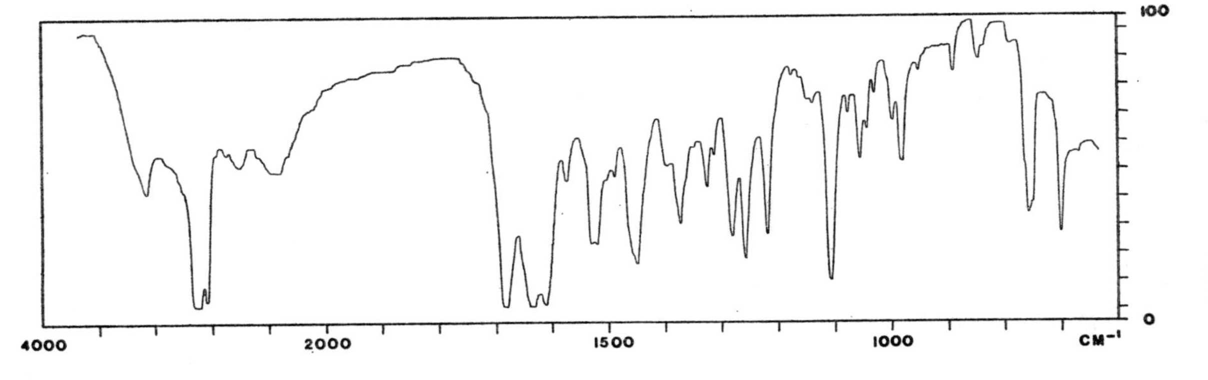

FIG. 7.68 (top). Phenmetrazine theoclate mull.

FIG. 7.69 (middle). Sample of Fig. 7.68 after standing overnight. A further example of crystal changes in a mull. Although of a very subtle nature, there are a number of bands of nearly equal intensity which could be chosen for indexing. The changes are sufficient to alter the whole picture.

FIG. 7.70 (bottom). A further mull of phenmetrazine theoclate to emphasize the point.

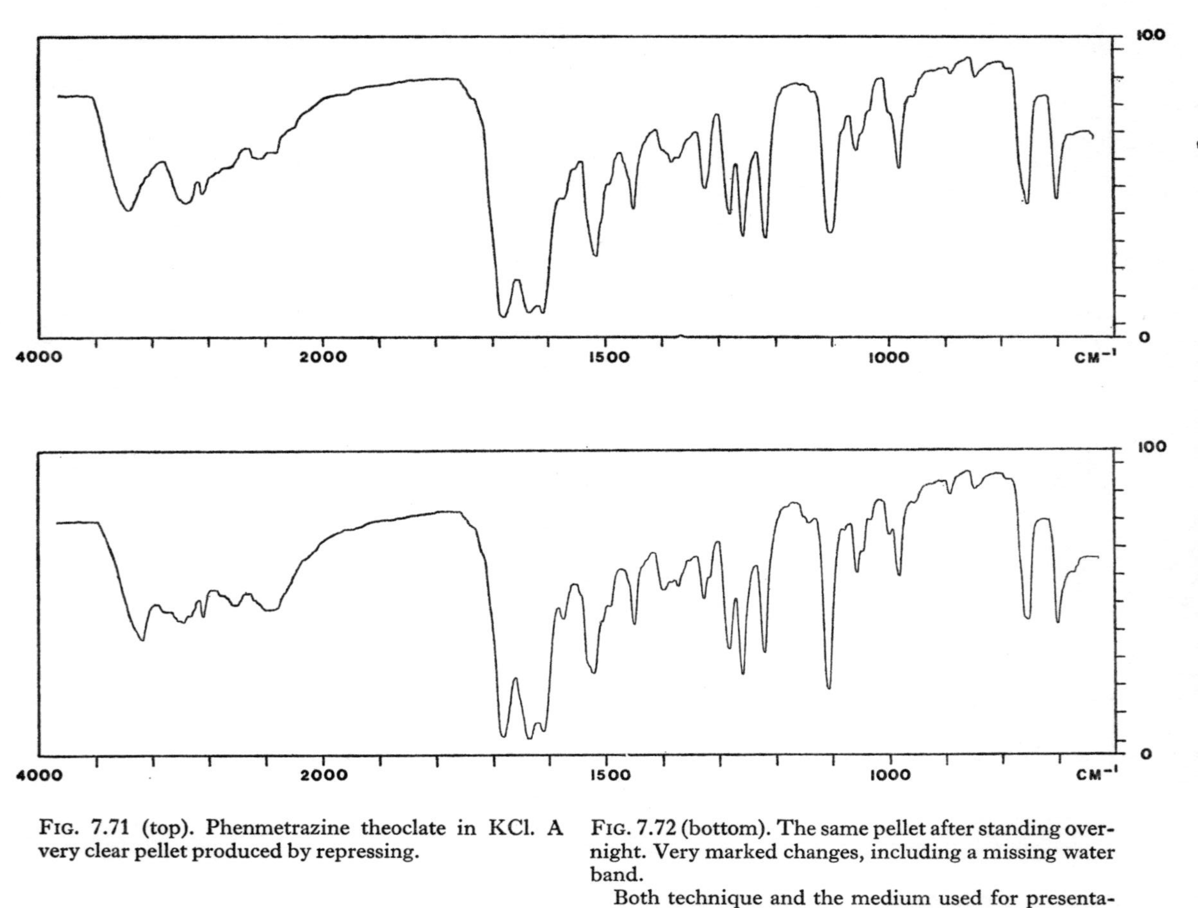

FIG. 7.71 (top). Phenmetrazine theoclate in KCl. A very clear pellet produced by repressing.

FIG. 7.72 (bottom). The same pellet after standing overnight. Very marked changes, including a missing water band.

Both technique and the medium used for presentation have an important bearing on the final spectrum.

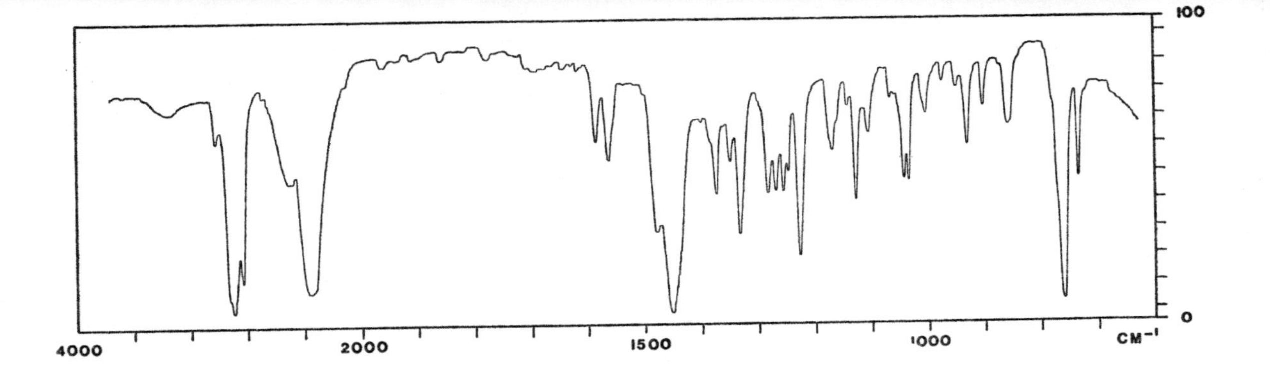

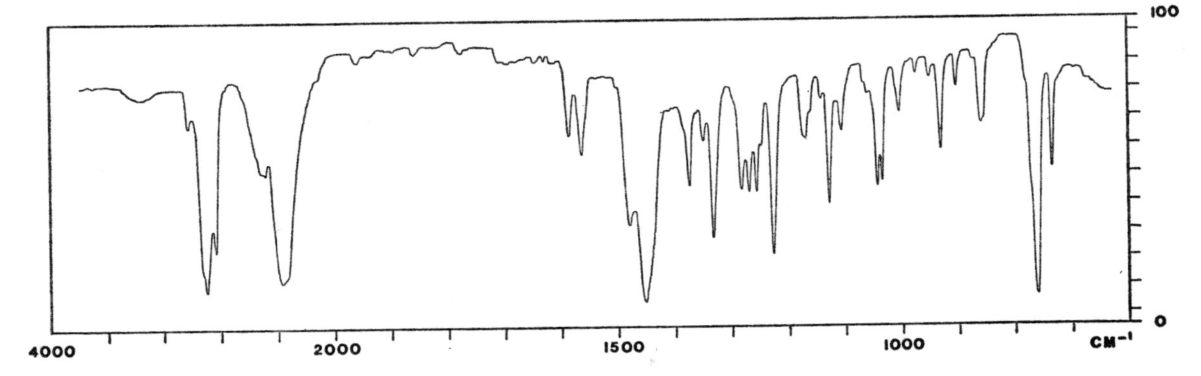

FIG. 7.73 (top). Promethazine hydrochloride mull. This sample is some years old and slightly discoloured, but the spectrum agrees well with the one of the American product in KCl which will be found in *J.A.O.A.C.*, 1962, **45**, 852.

FIG. 7.74 (bottom). The mull of Fig. 7.73 left for one week. The slight changes in relative intensities make all the difference to the indexing.

When considering a spectrum with many bands of nearly equal intensity, do not take a solitary example for indexing but investigate the effect of a wide range of preparation, even in mulls.

Derivatives (Figs. 7·75–7·76)

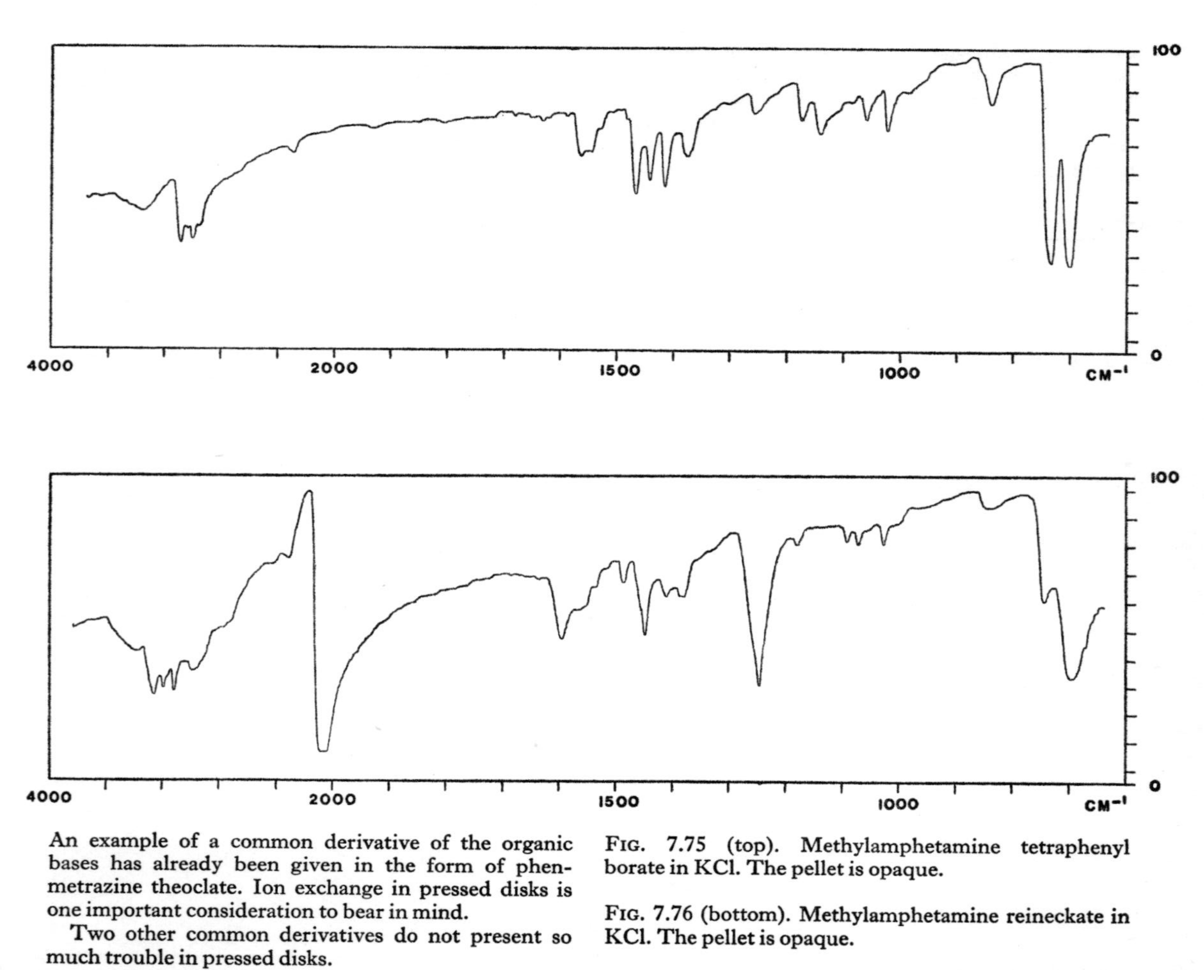

An example of a common derivative of the organic bases has already been given in the form of phenmetrazine theoclate. Ion exchange in pressed disks is one important consideration to bear in mind.

Two other common derivatives do not present so much trouble in pressed disks.

Fig. 7.75 (top). Methylamphetamine tetraphenyl borate in KCl. The pellet is opaque.

Fig. 7.76 (bottom). Methylamphetamine reineckate in KCl. The pellet is opaque.

8

INFORMATION RETRIEVAL

COMMERCIAL VENTURES

> Once I guessed right, and I got credit by't; thrice I guessed wrong and I kept my credit on.
>
> JONATHAN SWIFT

Information retrieval is very much a vexed problem. Finance is a prime consideration in reaching a decision on data storage and retrieval systems. Large organizations and, of course, committees can call on considerable resources and will attempt the more comprehensive schemes covering most eventualities, whilst the average spectroscopist through sheer necessity will work on a more hand-to-mouth basis, and rely on a do-it-yourself scheme. Both approaches fail because of their inflexibility or inability to cope with the volume of data available. At the same time there would appear to be a general non-critical data appraisal, leading to the acceptance of very poor spectra. Commercial data systems tend not to fall into this pitfall by their policy of selecting the best spectra to hand, or by acknowledging the source and leaving the reader to make his own appraisal; one bad spectrum is probably better than none at all. Consequently, laboratories which base their schemes on one of the commercially available indexes are in a much better position, since this does not preclude them from making other arrangements at the same time.

The following discussion gives a very brief sketch of the commercial indexes and a critique of other proposals, and is followed by a few observations for the true do-it-yourself addict.

COMMERCIAL INDEXES

Only two commercial indexes covering the bulk of the literature are produced on a regular basis in a form suitable for immediate use. Comprehensive descriptive literature about them can be obtained from their publishers, and complete sets are maintained in all the major reference libraries. All but a small fraction of the spectra contained in these indexes are of no relevance at all to forensic laboratories, and of limited value to the majority of research establishments. Nevertheless, the intrinsic value is incalculable, and over a period of time it is inevitable that all spectra will be of interest to someone—otherwise there is no point in the investigator publishing the spectrum in the first place.

Each index has its own approach to information storage and retrieval.

Sadtler Catalogue

The *Sadtler Catalogue* is by far the largest single collection of actual spectra, and increases by about 2000 spectra a year.[168] Spectra linear in wavelength are presented for the range 2–15 μm in both the pure compound and 'commercial spectra' sections. New additions are incorporated into five annual index supplements:

1. *Alphabetical Index*. The nomenclature of *Chemical Abstracts* is used and can lead to difficulties in the chemical interpretation of spectra recorded before the mid-1960's owing to the change in nomenclature over the years. Master the language and then no further problems will arise provided a copy of *Chemical Abstracts* is used to act as a referee in disputes over nomenclature. A few of the more common names are included.

2. *Molecular Formula Index*. Listing is in increasing order of carbon, hydrogen and other elements. Nearly all materials are listed here and in many respects this is a more useful index than the alphabetical one in view of the gradual adoption of new, and different, approved names.

3. *Chemical Classes Index.* This index is based on the *Chemical Abstracts*' functional groups order of precedence, the first three functional groups only being listed, except in special cases such as the steroids. Provided some clue can be gleaned from the unknown spectrum, this index may help to select a small group of spectra capable of being scanned visually for matches.

4. *Numerical Index.* All spectra reproduced in the main index are listed in numerical order, a most valuable feature. This index is invaluable where the spectra are not immediately to hand, because retrieval schemes usually produce a spectral index number only. Illogical retrievals are eliminated at this stage; for example, when a substance is known to belong to one chemical class and the indicated substance to another. Alternatively, the reference spectrum is not recorded in the *Sadtler Catalogue*. Reference to the molecular formula index should indicate which alternative is the more probable.

5. *Commercial Index.* This is split into two lists covering the commercial spectra catalogue. The alphabetical index provided for this section includes a number of trade names which need some other index for translation into a more familiar chemical description.

Specfinder

Any combination of the five previous indexes can be used to retrieve spectra when the relevant information is known. For a complete unknown, the Specfinder, or sixth index, is needed. Spectra are listed in order of the strongest single band wavelength, at intervals of 0·1 μm, with the remainder of the spectrum coded on the basis of the strongest band in each micrometre interval, taken to the nearest 0·1 μm.

The scheme relies on an accurate identification of the strongest band, and on accurate wavelength calibration. The absence of a strong band in any interval is a very important feature, always assuming that the spectrum has been recorded correctly, or at least under the same conditions.

DMS Cards

Spectra are reproduced to a standard size on an edge-punched index card, and are linear in wavenumber with the usual fourfold scale compression between 4000 and 2000 cm^{-1}.[3,63,189] The bottom and right-hand edges are punched according to frequency and intensity of main bands, and the other edges with details of chemical structure and a host of other physical parameters.

There is no alphabetical index and reliance is placed on a combination of a molecular formula index, chemical class, and spectral characteristics. In common with all edge-punched card systems, it is difficult to sort manually unless well subdivided. An optical coincidence index is available.

Separate literature cards are offered and can be quite useful.

ASTM Cards

Also in this group of commercial indexes, but operating on a slightly different principle, is the Wyandotte–ASTM body-punched card index.[165]

Cards are punched with a wide variety of information. Spectra are coded on the basis of 0·1 μm intervals, with all bands with an absorbance greater than 10 per cent of that of the most intense band being recorded. This can lead to difficulties when a negative sort is attempted, because slight variations in sampling technique can lead to a band being recorded, when, in normal circumstances, it would not. Structural units, similar in principle to the DMS scheme, are punched, together with the molecular formula, physical data, and the literature source.

Access to an IBM sorter, or its equivalent, is an essential requisite, and even then a spectrum is not retrieved, only details of it. It takes approximately 4 hours to sort all 100 000 cards with an average 6–7 card/second sorter, for each eight character search. Several hundred cards may be isolated on the first search, and multiple searches are needed. These subsequent ones need not be for a particular character's presence, rather for its absence—a negative sort. These negative sorts can only proceed on one characteristic at a time without more sophisticated sorters, and, as indicated above, they may be abortive.

punched-card index. Four hundred and fifty spectra are given and several times this number would be manageable if the index is sub-divided sensibly. Alternatively, a feature card index would be more compact and rapid in operation.

The book was critically reviewed.[96] Criticism was centred, as far as the infra-red section is concerned, on the supposed necessity of reproducing all the spectra and on potential errors arising from (in particular) polymorphism. While the book contains a number of spectra not readily accessible elsewhere in the literature, its serious drawback is the omission of such mundane materials as amphetamine salts and diamorphine. The smallness of the spectra reproduced was also criticized, but the overall pattern seen in such small and compact spectra is, if anything, more important than an accurate measurement of individual band positions, which can be derived from the full-scale original and itemized in the index.

Unfortunately quite a number of Clarke's spectra display one or more of the technical faults outlined in Chapter 1. Sulphates are very bad candidates for the pressed-disk technique because of their prevalence for ion exchange and the poor guarantee of the exchange's proceeding to completion. Clarke gives a spectrum of dexamphetamine sulphate, which, in my opinion, is that of dexamphetamine hydrobromide mixed with K_2SO_4, the single sulphate band at 1118 cm^{-1} being precisely the same as the sulphate band from a KBr disk of K_2SO_4. The base-line of his spectrum slopes badly and consequently the very strong benzene monosubstitution bands between 800 and 650 cm^{-1} are not indexed. Good examples of sulphate spectra are the A.O.A.C. spectrum of racemic amphetamine sulphate in KBr,[89] which shows no evidence of base-line drift or ion exchange, and the spectrum of dexamphetamine sulphate,[184] which has little evidence of ion exchange. The A.O.A.C. remarks on the use of KBr in place of KCl for sulphates are extremely relevant.[88]

The Clarke approach is basically sound for collections of up to 5000 spectra and only fails, in this instance, through a non-critical appraisal of the starting material. The most common fault noted was evidence of inadequate grinding. If pure substances prepared under supposedly ideal conditions are not positively identified, one can hardly expect items recovered from biological samples to be retrieved with any degree of confidence.

Optical Coincidence Cards

The most comprehensive scheme was proposed in 1963 by Schlichter and Wallace who suggested using special commercial cards.[174] Two years previously, Zink had suggested the use of stock inventory cards for the same purpose.[210] In its more modern form—the scheme of A. S. Curry—cards are produced lithographically from master cards, and the master spectral index recorded on microfilm.[51] The whole retrieval system is therefore in a convenient form for widespread distribution.

Unlike the Clarke system, no attempt is made to determine relative intensities, and six major bands are chosen, not three. This gives the scheme some degree of robustness. Bands are chosen at 0·1 μm intervals from 5 to 15 μm, except in the 6·7 to 7·6 μm range to avoid complications from paraffin oil mulls. As in Clarke's system, the same criterion is chosen for selecting major bands, with a specific note (p. 277) that . . . 'we have defined the strongest bands as those bands whose peaks are nearest to 0 per cent transmittance, *irrespective of the slope of the background* . . .' (my italics). Allowance is made for experimental conditions by coding bands with a tolerance of ±0·1 μm. Retrieval is extremely rapid, requiring selection of a maximum of six cards, an answer being possible with less than this. Unique answers are not always possible, even when due allowance is made for spurious transmission, but the number of potential answers produced is small enough for simple manual sorting of original spectra. Occasionally a nil answer is returned, indicating that the spectrum is not recorded in the master index, or that slight errors in wavelength measurement have occurred, or that the master card was incorrectly punched. The lithographic reproduction then comes into its own by showing 'near misses' as regions of relatively high light transmission in the black background.

The main advantages of A. S. Curry's scheme are its versatility, simplicity in use, and high speed.[51] Extra diagnostic features can be added at any time, and recovery speed surpasses a computer by the ability of the operator to alter his programme as he progresses with the search. The cards used by Curry could accommodate 3600 items, a limit well within the capacity of his reference library; a card of double capacity is being developed.

Failures are attributed to four factors:

"1. Impurities . . . may introduce additional bands . . .
2. Variations in the levels of background absorptions . . .
3. Polymorphism . . .
4. Grating instruments with high resolution may resolve peaks which appear as shoulders with instruments incorporating only prism optical systems.

It will be obvious that these factors will cause difficulties with all peak position infra-red retrieval systems."

As a basis for data retrieval, the Curry system is excellent, although the original had a few transcription errors, and is fundamentally the one adopted later in this book. No effort is required in setting up the index since it is prepared centrally and distributed. Setting up a similar scheme oneself can be tedious and is discussed later (p. 182).

MIRACODE

Complete details of the MIRACODE are given by Thomas[186] and are not relevant here. The scheme is based on a commercially available data scanning system having some similarities to Curry's proposals. It takes the ideas one stage further and places all the retrieval information on the same microfilm as the spectral index, which is then scanned by a high speed photoelectric device.

The high installation costs and the relatively long periods of inactivity render this system of academic interest only to all but the larger organizations (where it is relatively inaccessible). However, it is by far the most comprehensive scheme proposed and may become more common in time. Its primary advantage, once the main console has been acquired, is its ability to accept any other request or specialist programmes fed into it from some central clearing agency. In the meantime, separate indexing of spectra and relevant data would appear to be the only alternative.

Edge-punched Cards

So far no mention has been made of edge-punched cards in this section. A perfectly good scheme is available commercially, namely DMS cards. The reader can choose to adopt this without comment, build on it by producing his own cards from DMS card blanks, devise his own system, or combine all three approaches. Other independent schemes have been proposed in the literature from time to time, the limit on these being placed by the time available for coding, or the ingenuity of the author, or both.

9

PRACTICAL SUGGESTIONS FOR INFORMATION RETRIEVAL

Nothing is given so ungrudgingly as advice.

LA ROCHE

It should be obvious by now that a data retrieval system must be sufficiently flexible to cater for a huge quantity of inferior information. Alternatively, only well-qualified staff should be allowed near any stage of infra-red preparation or indexing. In view of the widespread errors found in the literature—almost a universal factor, in fact—it seems safest to devise an index/retrieval system in which errors of translation etc. can be detected, rather than hope that all operators will reach a satisfactory level of competence. This means that the volume of material requiring indexing will increase severalfold not decrease, and the system must be sufficiently flexible to incorporate new parameters and a continuous flow of new material. An index is only as good as its starting material,[52] which means in practice the operator, and to be of general value it must incorporate *all* facts known, and not be selective or subjective.

Computer-based schemes and the MIRACODE have reached a degree of sophistication and reliability more than adequate for the purpose in hand and one will undoubtedly be universally adopted in time. Meanwhile, the choice remains between one of the card index approaches. Nearly all these mean two or more parallel indexes. To be really effective, the following information must be provided:

1. Indexes from which a spectrum can be retrieved from non-spectroscopic data, and

2. An index from which a spectrum can be recovered, by its spectral characteristics, for comparison with that of an unknown.

These two main requirements do not mix well and are most conveniently treated separately.

RECOVERY FROM NON-SPECTROSCOPIC DATA

The bulk of spectra consulted arise from simple questions of the type—'Do we have a spectrum of X?' 'How do spectra of Y compounds compare with X compounds?' A fair proportion of workers will never need to answer more than these simple questions or will be content to stop with this type of retrieval. A simple card index will usually suffice; a sophisticated sorting system is not necessary.

Four indexes are needed for a complete coverage of all eventualities, but, of these, only one is an absolute necessity. All four indexes are provided free to *Sadtler Catalogue* subscribers, although these will be of limited value in specialized work on compounds not covered by the *Sadtler Catalogue*. It is assumed that the reader wishes to set up his own scheme, based on these four logical indexes. In the course of time, some extension of the Sadtler coverage will be necessary, in any event, to cater for the unique spectra covered by the reader's own field of study. The indexes are:

1. *Alphabetical index of all compounds with recorded spectra.* As a minimum, compounds must be indexed alphabetically by chemical name as well as by common name, where this is approved, and, ideally, all trade names should be included. Indexing trade names is an undertaking of no small magnitude in itself, and is not necessary if the information can be obtained from another source, such as 'Martindale' or the *Merck Index*. Some economy can be made by utilizing the reference sample index to record this information. A single card per substance is needed for maximum flexibility.

2. *Numerical index.* This is necessary where the original spec-

trum is not immediately to hand. The A. S. Curry scheme would then be usable in situations where the microfilm reader/printer is not installed, as the worker could then consult his own spectra of the materials indicated. Visual scanning of the alphabetical index provided is time consuming (on over 3000 compounds it is quite unreasonable) and, at the same time, the vital number may be missed, necessitating a multiple search. This numerical index is best kept in a log book or on typed sheets for ease of reference.

3. *Molecular formula index.* This is invaluable when differences occur in the nomenclature used in the general index. In the present context, the alphabetical index is already a distillation of the required information and makes a molecular formula index redundant. The molecular formula is rarely available for a true unknown.

4. *Chemical class index.* This tends to be built into one of the other indexes because of uncertainties with an unknown. Simple classifications such as acid/base/neutral have been elucidated by the extraction process and are adequate to eliminate nonsensical answers thrown up by the primary retrieval system. The intricate detail shown on a DMS card is not necessary.

Indexes 1 and 2 are straightforward and common to all envisaged retrieval systems, and indexes 3 and 4 can safely be ignored when the work in preparing them is considered and related to the return.

RECOVERY FROM SPECTROSCOPIC DATA

A solution to the problem of identifying a spectrum rests to a large extent on what the system is intended to achieve. The larger commercial indexes cover nearly all known spectra and the bulk of their cards are therefore never consulted. By being selective, almost any scheme is a practical proposition as the total number of spectra envisaged will not reach 10 000 for some time, unless all commercial mixtures (e.g. tablets) are incorporated. The relative merits of edge-punched and feature cards have been extensively discussed in the literature and each has its advocates.

D. R. Curry's review is a concise summary of the relative merits.[52] The paper by Anderson and Covert[2] is also appropriate. If it is intended to retain all original spectra and merely use the retrieval system to indicate those of interest, the feature card approach is undoubtably the quickest and most adaptable. The original spectra are stored in numerical order in files, or recorded in sequence on microfilm as proposed by A. S. Curry.[51] Edge-punched cards become competitive when all the relevant data, including spectra, are recorded on the cards. A DMS card recording facility on the spectrometer is not an essential requisite.

Selecting Spectral Characteristics

It is generally agreed that, for spectrum identification, a minimum of five to six bands must be selected to eliminate ambiguities, but, for the individual worker in isolation, Clarke's scheme based on three bands is more than adequate. This is effectively six parameters. The total number of spectra need only be around a thousand—500 pure substances, each in a pressed disk and a mull, with a few in solution or thin films. An ordinary plain-card index can cope with several times this number of entries.

Where collaboration between laboratories is envisaged, the work of Thomas and A. S. Curry noted above has shown that coding on six bands is a convenient criterion. The decision then rests on the method of selection. If this is based on bands with peaks closest to the 100 per cent absorption abscissa, it requires no skill, accurate measurement, or critical assessment, and is therefore ideally suited for use by technical assistants. It will also guarantee that all spectra, regardless of quality, are incorporated into the index on an equal footing; selection can take place later by the person actually using the retrieved spectra.

Both Sadtler and A. S. Curry divide their spectra into 0·1 μm intervals. This is an obsolete approach in view of the swing towards linear wavenumbers. If the range 2000 cm^{-1} to 650 cm^{-1} is agreed to be the most useful portion of a spectrum for indexing, division into 10 cm^{-1} intervals is a logical alternative. This yields a total of 135 features instead of 100, an apparently finer subdivision. However, few spectra have bands between 2000 and 1800 and beyond 670 cm^{-1}, and the comparison is now effectively 113 to 94. If the 6·7–7·6 μm range is ignored as advocated by A. S. Curry, the number of intervals available becomes very similar at approximately 94 to 86. Therefore, there is little difference between either alternative as far as accuracy is concerned, but the linear wavenumber presentation is at a distinct advantage at long wavelengths because of the apparent sharpening of bands, making assignment to a particular interval that much easier.

Division into 25 cm^{-1} intervals between 700 and 1000 cm^{-1} and 50 cm^{-1} intervals between 1000 and 1800 cm^{-1} follows the DMS lead and renders it more amenable to edge-punched-card indexing by reducing the number of intervals to 28 or 30 if a 'dustbin' is added at each end. Bands will tend to be clearly in or out of a particular interval, and, in cases of doubt, only two need be punched. Standard 8 × 5 inch edge-punched cards are available with 33 holes along the top edge, or an even closer spacing with 41 holes, ample to index the whole spectrum from 650 to 4000 cm^{-1} in DMS intervals.

Checking Spectral Characteristics

Once the decision has been taken on the criteria for indexing bands, it is essential to consider carefully the best method of measuring bands for this indexing and, subsequently, for checking the retrieved spectrum against that of the spectrum being analysed.

Band measurement

It is rare for commercial charts to be printed without a grid pattern for measurement of both frequency and absorption, and, for routine use within any single laboratory, this grid is ideal. The manufacturer has, after all, taken the trouble to provide the most useful general-purpose grid. Problems arise when the grid is the wrong one for the index scheme (wavenumber for wavelength, for instance), particularly when spectra from the literature are being processed. A number of workers surmount this problem by visual interpolation, with or without the aid of a ruler. On a routine basis this is tedious and can lead to subjective errors. For a small number of measurements to moderate accuracy, visual interpolation is recommended.

There are two modifications of a simple expedient—the use of light—in wide vogue. When the charts are thin, a blank chart is superimposed on the thin chart and both are held up to the light and the appropriate measurements are made. When the charts are thick (e.g. DMS cards) or unwieldy, or simply for practical convenience, a light-box is preferred. The method of use with either variant is the same, and is discussed next and also in the section on spectral comparison (p. 177).

Light-box work is impractical for published spectra—there is usually printing on the reverse page—and for spectra traced in pale ink or in ink of a similar colour to the grid. The following method is recommended as a general method for band measurement. Take a Xerox of a blank chart or of an example from the literature on a polyester sheet and use this Xerox as a transparent plan. The top or bottom of the grid pattern will be marked off in the alternative units to the main grid. As the transparent plan is moved up and down, each band's peak will coincide in turn with the required scale and can be read off directly without eye strain or error. This process will also read off at the same time the band intensities in their correct order without subjective errors arising from false alignments by eye. If the required frequency scale is at the bottom of the grid, it may be found more logical to guillotine the grid and use the remains as an ordinary ruler. Should this method be chosen, remember it is easy to cant the ruler, in spite of the horizontal rulings, because the vernier effect with the grid is now absent; the risk of doing this is greater when measuring spectra reproduced 'clean' as in this book.

Any slight errors arising from a reproduction of a different size to that of the Xerox can be annoying but are rarely severe enough to upset the relatively crude measurements needed for a retrieval scheme. Adjust the Xerox to obtain the reproduction the correct size.

It must be admitted that this technique is not very practical for the very large charts used by some spectrometers, even when strips are Xeroxed and joined, and the analyst will be forced to adopt one of the other (conventional) approaches. A large chart is easier to measure anyway, and subjective errors should be less prevalent.

Spectral comparison

The approach adopted when comparing spectra depends to a large extent on the attitude of the individual. The purist will insist on large-scale reproduction and a precise matching of each band in the two spectra; this matching is easiest with a light box. The practical person will be more concerned with a matching of patterns and the non absence (not just the presence) of certain other diagnostic bands and patterns; small-scale reproduction is often found more convenient for this. Once the pattern is recognized, a quick visual check for foreign bands, and some sample band frequencies, will suffice to confirm the identity. A more detailed examination is only justified when anomalous spectra are encountered, for instance when the spectrum is the result of co-extracted interference.

The simple method of direct examination is to superimpose the two spectra and hold them up to the light. The better method is to superimpose spectra over a light box. In either event, points of variance stand out immediately and should be ticked off (there should be less of these than points of similarity) and studied for relevance. Maladjusted charts are also picked up very quickly when spectra are juggled to superimpose as much as possible of the two tracings, with due allowance for base-line drift, of course.

It has always puzzled me why some workers always insist on placing the comparison spectrum over the reference one before checking on a light box. Always place the spectrum with the more opaque ink underneath.

Edge-punched Cards

Most modern spectrometers have a built-in facility for spectra of the DMS size, or an 'optional' extra is offered. Some spectrometers can be fitted with an auxiliary recorder,[164] whilst the alternative is a pantographic reduction.[194] The main advantage of recording spectra at DMS size is subsequent ease of storage, regardless of whether or not they are incorporated into a retrieval scheme. Originals at full size can be retained or not as desired, but if destroyed, all accurate measurements of features used for indexing must be recorded from the full-sized chart and noted somewhere, preferably with the reproduction, as with DMS cards. Several small reproductions can be taken-in at a glance and an overall pattern assessed before intricate details are considered.

1. *DMS cards.* DMS cards are readily available and are familiar. No thought is necessary to adapt them to the commercial index, full instructions being supplied with the original set. The card's primary index is chemical class (punched along the top) and is largely redundant in a specialized index where chemical structure is often unknown.

It takes a long time and some chemical knowledge to make an accurate index based on chemical class, and is left well alone, in which case, all DMS cards are best stored and used upside down so that the prime index (along the new top) now becomes the spectral range 3200 to 1100 cm^{-1}. Blank cards are relatively expensive, are not a familiar standard size being A8, although they will store conveniently in an 8 × 5 inch filing cabinet, and if used as suggested, half the indexing capacity is wasted. If spectra are printed in black, a very bright light-box will allow two cards to be superimposed for direct visual comparison.

2. *Adapted cards.* Edge-punched cards are widely used in other aspects of chemical work and the same cards can be used for infra-red retrieval simply by changing the coding. The tablet identification cards of McArdle and Skew[127] provide 32 spaces along the top, just sufficient for the DMS notation. For real economy, the cards could be left unaltered with slotting and reading-off simplified by reference to some master card drawn up for the occasion. Alternatively, the cards are relatively 'clean' on the back and could be used back-to-front for ease of reading.

The only written information needed on the card is the serial index number from the spectral index and the name of the substance; all other information ought to be on the original chart. Any other card can be overprinted with this small amount of information without excessive hardship.[126]

The idea of using what amounts to second-hand cards in this manner is not aesthetically pleasing and does not appear to have been adopted to any extent, but should one of the edges of another index's card be unused for recording information on the original subject, this edge may be used for infra-red retrieval, thereby making the cards dual-purpose. An obvious example is the index for pure compounds in a reference collection, where the same card would now give spectral information as well as physical data. Even more pertinent would be the pressing into service of a spare edge on a tablet collection card by which the tablet's spectrum (derived from some formalized technique, such as a chloroform smear on rock salt) could be recovered (Plate 13). Some tablets are badly mutilated on receipt (powdered, damaged, or partially digested) and the normal visual characterization of the tablet may not be possible. The utilization of a special self-contained index for tablet/capsule identification, divorced from the main infra-red index, is worthy of serious consideration.

3. *Self-designed cards.* Offset litho and allied printing methods are now so universal and cheap in application that there is little point is using adapted cards unless they are utilized in conjunction with another index (see above). Even then, it may pay to redesign them to a specific need and print a supply. Once the master block is produced, fresh batches of cards can be printed almost anywhere at a fraction of the cost of commercially designed cards purchased through normal channels.

The master design can include anything of value. In its simplest form it will be no more than the hole/punch code and space for the serial number and name of compound, printed on one side only, whilst a more sophisticated version will look more like a DMS card, complete with spectrum, with printing on both sides. This latter point is an overriding consideration in deciding to produce one's own cards. Litho reproduction is so precise that one is not restricted to using just DMS size charts on the card, and a completely redesigned full-sized chart for the spectrometer in question, complete with a built-in retrieval system, is a very practical proposition.

To take specific examples, normal chart paper for a Perkin-Elmer 137 spectrometer is not a standard edge-punched card size, but specialist firms will quote accordingly and for the necessary litho printing.[8] Ordinary thin card will roll around the chart drum without difficulty and a spectrum recorded in the conventional manner. If the spectrum is worth adding to the index, the appropriate slots can be cut in the card's periphery, and the result filed in the same way as a DMS card. The charts are essentially A4 size and are conveniently stored in an A4 cabinet, or a foolscap cabinet (with lateral padding if desired). Charts for a Unicam SP200 spectrometer or the Infrascan are even simpler as these can just be squeezed onto a standard $11\frac{3}{4} \times 8\frac{1}{4}$ inch card, and with a magnetic clamp of the type used on the Infrascan, any lateral adjustments needed are easily accomplished. However, charts of this size are a little cumbersome in use and can be damaged with rough handling, as well as causing storage problems. Since retrieval is invariably based on the spectra range 2000–700 cm^{-1}, only this portion need be reproduced on the cards. Unicam charts will now go on a standard $8\frac{1}{4} \times 8\frac{1}{4}$ inch card, whilst the middle sized Infrascan charts (twice DMS size) will just go on a DMS card ($8\frac{1}{4} \times 5\frac{3}{4}$ inches) if the side holes are sacrificed. Most other spectrometers can be treated in a similar way with a little ingenuity; with a little thought, the edge-punched hole's spacing can be made to correspond to the hole spacing along the charts for 'continuous flow' recorders.

Considerable economies in storage space are possible if DMS chart facilities are available. The 2000–650 cm^{-1} range will go on to a 5×3 inch card with ample room to spare, and with 25 holes provided along the top edge, a workable retrieval scheme can be devised without recourse to the other edges. One of the tablet identification schemes relies on a 6×4 inch card[126] and if the numerical index along the bottom is sacrificed, this could be used for the infra-red index, with the spectrum printed on the back (upside down, relative to the front, for convenience). The extra printing required is carried out on an otherwise blank or wasted back of another card and is an extension of the principle of adapting an existing card index. The possibilities for properly designed cards tailored to one's own need are endless. The cards do not need to be pre-punched and ordinary plain cards can be slotted as required.[4] If an edge-punched-card index system is chosen, it will become damaged and unwieldy in use if all the cards are sorted each time a search is made, even if one of the commercial sorting frames is utilized. These frames, or hand sorting by needles, are capable of handling only a few hundred cards at a time, and a complete search becomes tedious. Some form of pre-sorting is advisable, each subdivision being within the capabilities of a single search. A decision on the type of pre-sorting is not easy and is best made on the basis of one's own peculiar requirements. Sorting into groups according to the strongest band is an obvious choice, but this makes no allowance for changes in relative intensities in, say, an extract where traces of impurity can have a marked effect.

Another shortcoming of punched cards bearing spectra is the necessity of holding parallel spectra for ease of retrieval from alphabetical indexes. The alternative is a search through all cards by a punched number index derived from an alphabetical index, all extra wear and tear. However, original spectra will usually be retained for accurate measurements and for direct visual comparisons, and the parallel collection is therefore already to hand. In view of the frequency with which errors are incorporated into an index, the retention of all original spectra would appear to be a wise precaution in any event.

Feature Cards

For widespread distribution, the scheme of A. S. Curry[51] is supreme. Extra sets are reproduced with little effort and at a proportionately lower cost. It does require capital expenditure on a microfilm reader/printer; in the absence of this, photocopies of the original spectra must be distributed, and this negates the savings in storage space and access time offered by the microfilm. Another drawback is the limited storage capacity of the feature cards described, because, to be really effective, the index must include a much larger collection of spectra. Three thousand, six hundred is just not enough, even if all of them were first-class quality, and the proposed development to a total capacity of 7000 units could not cope with a single example of all common pharmaceuticals, in their normal forms from each manufacturer, in mull, pressed disk and solution, let alone start to consider polymorphs, typical extracts, or other salts.

The answer would appear to lie in running parallel feature card indexes, each one devoted to a specific topic. All mull spectra could appear in the first one, numbered 1 to 7000, all pressed disks in the second set, numbered 7001 to 14 000, and so on. The total capacity would then probably suffice to cover all eventualities. Needless to say, very little of this information would ever be consulted, fewer than 100 pharmaceuticals and poisons being encountered with any regularity, but there should be ample capacity to cater for new materials coming onto the market. Also, the scheme is proposed for wide distribution and must cope with a wide variety of expertise, equipment and local problems.

Expansion of card capacity to 10 000 units, as in the Schlichter and Wallace scheme,[174] gives a more obvious meaning to the subdivided numerical index, since 7000 is not very convenient. Feature cards of 10 000 unit capacity are commercially available.[118]

Further complications arise in the different resolution of the common spectrometers. Ideally, all spectra should be compared on the same type of instrument, but this is never possible in practice, if only for the obvious reason that each laboratory has a different budget. Some allowance can be made by making separate feature cards listing spectra linear in wavenumber or wavelength, and in an ideal situation, separate cards for each type of spectrometer, but the return on the time and effort in completing them would be marginal. A split index of this type could be useful in situations where a number of approximately equal intensity bands occur which are given a differing relative emphasis by each make or type of spectrometer.

Setting up a Feature Card Index

General comments

Until a really comprehensive commercial or collaborative feature card index becomes a reality, and the errors in them eliminated, the reader is probably better off manufacturing his own index. In order to make full use of the rapid retrieval possible from feature cards, the index should not be restricted to spectra produced solely in the operator's own laboratory. Every source, including published atlases, needs culling, and in this way a much more comprehensive pattern of errors and differing spectrometer biases will be built into the index. In the course of time, the percentage 'hits' will rise to 100 per cent, or a close approximation, on each occasion that the index is consulted because all the variables will be increasingly encompassed. Ninety-six per cent efficiency on first search, as cited by A. S. Curry, is not unexpected considering that the same instruments, or similar instruments in good repair, and the same charts, or staff of an equivalent level of competence, were employed in the exercise. Recovery efficiency from varied instruments, possibly in poor repair, and staff with a variety of skills and experience can be lower than 50 per cent, or an even chance of making no recovery at all. If the errors introduced into the master index are taken into consideration, the acquisition of such a centrally produced index can be more of a liability than an asset.

The above suggestions for a scheme for wide distribution are a minimum requirement and beyond the capacity of an average laboratory devoted to routine work and earning an honest living. The best solution for an average laboratory lies in the staff making their own index based on their own work and as much published data as comes to their notice. The analytical methods used will, of necessity, be well standardized and problems of polymorphism and the like are automatically surmounted by incorporating into the index all spectra *NOT already retrieved* from the existing index. Theoretical consideration of the reasons why any particular spectrum is sub-standard is irrelevant, since the important criterion is that whatever went wrong (and it may simply be material from a different source with different impurities) it has happened once and can easily happen again. It is therefore a vitally important fact to build into any index. Indexes relying on the presentation of a single good spectrum do not live in a real world.

The storage capacity of a feature card index is regarded as a limiting factor for convenient use. This need not be true if some prior thought is given to subdivision into parallel indexes as indicated above. The suggestion of index division into mulls, pressed disks, liquids (including solutions), tablet smears, and other techniques gives a total capacity of 50 000 spectra, ample for all foreseeable needs. Each index can be used independently, without the necessity of doing five separate searches as one would need to when the total storage capacity reaches 50 000. Ten thousand mull spectra will take some considerable time to amass, and by the time this has been done it is to be hoped that collaborative schemes based on the MIRACODE on computer sorting will have come into their own and can be used in conjunction with the laboratory-made index. However, some form of local index will always be needed to cater for particular conditions and peculiarities. Further local index subdivision into the main analytical classes of acids, bases and neutral compounds is always open to consideration, although the dividing line between them is often nebulous. Cross checking an index is an essential requisite and must be carried out at all stages, particularly to check that the correct data are being selected; the problems of correcting a feature card are much more serious than with an edge-punched card.

Some spectra will be identical, as far as the retrieval scheme is concerned, regardless of phase presentation. Thus in cases where the mull spectrum is sufficiently similar in its gross detail to that of a pressed disk, the numerical index could cross-refer to the more useful spectrum (usually a pressed disk) and thereby save storing both originals. Although it could be regarded as an extra complication with parallel indexes, the advantages of a double or multiple check are important, and if it is decided to retain all originals, doubts on the importance of minor bands can be allayed expeditiously. In the conventional approach, with a single, non subdivided index, there is a temptation to economize and only enter one example, usually a pressed disk, and time is wasted running other reference spectra when minor bands are important; as has been discussed in the section on techniques, minor bands, and often major bands, are very critically dependent on personal expertise. Perhaps the simplest answer for a private index is to run all spectra in one phase at all times.

Duplicating feature cards

Hand duplication of feature cards is soul destroying. It may be necessary if errors are detected, or cards damaged, and professional reproduction by lithography for one or two copies is not economically justifiable unless the true cost of labour (and checking it) is taken into consideration. Xerox and other duplicating processes are capable of printing on polyester sheets. They do not produce a very good reproduction of large expanses of black from the average coloured feature card, but some control is possible from the contrast or 'colour' setting on the duplicator. This graininess has most of the desirable aspects of the cards on transparent sheet advocated by A. S. Curry, and at a fraction of the cost for small runs. Contrast can be enhanced by using black master cards, with a sheet of white paper placed over the reverse before duplication; the white paper sharpens the punched holes' edges by minimizing reflections and shadows. These duplicated cards are as robust as the commercial items and just as accurately duplicated, and may be accurately trimmed to size in an ordinary stationery guillotine. Provided the Xerox duplicator has been correctly adjusted, the reproductions will be precisely the same size as the originals and non-duplicated plain cards (punched to some local requirement) can be mixed in if desired. Normally, however, there is some 'barrel' distortion in the optics of the duplicator and the result is a slight distortion of the original proportions, in addition to a slight change of size. The reader is advised to test this factor before mixing in unduplicated cards. In either event, the cards will be the same size for storage purposes, and the same storage cabinet and light-box can be used without modification. Because the cards are not significantly reduced in size by this process, small edge-trimming errors are relatively unimportant and balance out any inconvenience in handling.

Adapting for other uses

It should be fairly obvious that feature cards may be adapted into or onto an existing system in the same way as has been discussed for edge-punched cards. Provided the principle of split feature card indexes is acceptable, the whole index may be expanded as each new topic proves useful.

As an example, it may be desirable to retrieve compounds on the basis of functional groups, very much as DMS cards are applied, without reference to infra-red spectrometry at all. A number or numbers will be indicated, which can be consulted in the spectral archive or in a separate numerical index. A separate numerical index of spectra would, therefore, be of some value to workers disinterested in infra-red work and wishing to avoid investing in a microfilm reader. Similarly, chromatographic data or physical constants may be built into the main index and used quite independently to indicate substances in another type of analysis. If the main infra-red index is split into 10 000 units, these subsidiary indexes need only be related to one of these units containing all compounds, say, pressed disks. Relatively few holes will be punched in these new cards since the spectral index will contain many examples from the same compound in differing guises.

There is one subdivision of the master index which is obviously tailor-made for adapting to other purposes. This is the tablet/capsule index found in laboratories concerned with the identification of poisons. With a little extra effort, the numbering of a tablet collection and the corresponding tablet/capsule spectra archive can be made to coincide, and the two sets of cards can be used independently or in conjunction as circumstances demand. Whilst an edge-punched-card index for tablets is almost universal, and can be combined with a duplicate spectral archive, the advantages of the flexibility offered by a feature card equivalent are enormous, particularly if it is consulted with any regularity. A separate alphabetical index for tablets is normally constructed automatically and, if based on a simple plain-card index, can carry any information not on the spectra, such as information on composition abstracted from a less obvious reference source.

10

INFORMATION RETRIEVAL

DATA APPRAISAL

Theory and practice always act upon one another.

GOETHE

Nearly all studies allow for errors in frequency measurement by incorporating a tolerance, which, in linear wavelength systems, is $\pm 0{\cdot}1$ μm as a rule. Very little attempt has been made in commercial studies to make an allowance for poor or anomalous spectra, and the ICI approach[49] is to be commended. Generally speaking, manually sorted punched cards are difficult to adapt into a scheme for assessing spectral quality or relative band intensities. Feature cards are easier to adapt but the time taken to make extra cards is an important consideration. The simplest compromise would be a separate card listing all spectra whose strongest band occurred at that particular wavelength, i.e. a 'Specfinder' approach. A. S. Curry found an average of 2·4 spectra per search in a population of 1100; with the extra criterion of the strongest band, the chances of a unique answer are greatly increased. Combining the Clarke approach with a feature card index is a logical conclusion for a really large population. In many cases, unique answers are possible long before all nine cards are utilized.

Some aspects of data appraisal are discussed below as an extension or application of factors discussed in the chapter on anomalous spectra (p. 64). In order of importance they are corrections, band assignment, and 'know thine enemy'.

CORRECTIONS

There are occasions when a poor spectrum is obtained from an unknown material and it is not possible to repeat or improve it. An illogical or incorrect answer from the index is a high probability in these circumstances, and a critical appraisal of the spectral features is an essential step in a cross check. Since nearly all proposed indexes rely on a complete disregard of the background or any base-line slope, and since reference samples will tend (it is hoped) to be correctly prepared, some allowance must be made for the background. Accurate allowances are obviously not possible, but a good guess is, and if the newly estimated band intensities make any difference to the indicated material, all candidates can be consulted in the master archive. The process can be repeated until a reference spectrum is found with the desired overall pattern, or defeat is admitted. When other preliminary chemical tests have indicated one material only, this data retrieval process is superfluous, the appropriate reference spectrum being produced for comparison, but there are always times when no prior data are available for assistance.

When the spectrum is positively identified, it should be entered into the index for future reference, *NOT* discarded.

Other corrections can be applied in the event of a technically faulty sample preparation. It is not easy to make accurate adjustments in situations where relative intensities have been altered, and next to impossible when the sample has undergone some transformation. Nevertheless, when the fault is correctly diagnosed, past experience is a useful guideline.

Allowances can be made for inadequate particulation, inadequate presentation, sample change, and interference.

Inadequate Particulation

1. *Broadening of bands.* Major bands are still obvious but their relative importance may be altered. The corrections are best applied on a trial by rule-of-thumb until some experience has been gained.

2. *Christiansen filter effect.* This tends to be more obvious at short wavelengths and does not always affect every band. Shifts in the frequency at the point of maximum absorption are normally within the tolerances built into the retrieval system, but bands near the edge of one interval could be carried into a far adjacent interval and not appear as a 'near miss'. In the first instance, only distorted bands should be considered for a wavelength correction.

3. *Base-line slope.* As with the Christiansen filter effect, the base-line slope is most obtrusive at short wavelengths, giving undue prominence to bands at this end of the spectrum. A simple adjustment is very convenient in practice, but only if the operator is familiar with the basic absorption laws and understands the interrelation between band width, intensity and wavelength. Base-line correction may have no effect on relative intensities when all factors are considered.

Unbalanced optics can lead to base-line slopes in either direction; unbalanced cells can give a variable slope in certain parts of the spectrum which may not be noticed. Unbalance is often noted at long wavelengths because of refractive index changes, lower infra-red energy and the sparseness of bands. The diagnostic mono-substituted benzene bands need close scrutiny when spectra are returned under conditions of high reference beam attenuations.

All three allowances are derived from the same basic cause and may be applied at one and the same time on any particular spectrum. They are too frequently needed to be dismissed as improbable, all the published atlases giving glaring examples of all three.

Inadequate Presentation

1. *Sample incompletely filling the beam.* This is most common with small samples which should have been properly masked and then compensated in the reference beam. Major bands are truncated. The fault is quite common with mulls when these are made too stiff and the two salt disks are slapped together without due care and attention.

2. *Sample too dense.* The major bands are off-scale, or too many of them seem to have an apparently equal intensity and importance. General rules for allowances are impossible—select combinations of bands and treat as a lottery. Fortunately, in general work, the shortage of sample is sufficient to make this fault a rarity in normal circumstances, but the temptation to take too much sample is very great when it does present itself.

When the sample is too dense, most operators take the line of least resistance and rotate the mull cell in the incident beam until some of the off-scale bands come back on scale. Conditions similar to 1 apply where the sample is incompletely filling the beam, and it is usual to find both faults occurring together when there is too much sample available for analysis. The remedy should be applied before recording a spectrum, namely, remake a thinner sample or improve the evenness of distribution.

3. *Spectrum crammed into lower portion of the chart.* This should only occur with the cheaper spectrometers not fitted with a sample beam trimmer or reference beam attenuator, but examples of this from instruments with all the basic accessories are very common in the literature. The visual appearance of the spectrum is similar to 1, except that it is in the high absorption portion of the chart instead of the low absorption region. It is a more serious fault than faults 1 and 2 because the truncation of bands affects all bands with a reasonable intensity, and distorts all the relative intensities.

Like fault 2, the spectrum can only be examined on the basis of experience, with clues provided by the band width and the overall pattern.

It should be noted that there may be a high energy loss from samples incompletely filling the beam (1 above) arising from the cell design or other factors. Thus, a mull cell may be made from two salt disks whose surfaces and spacing are such that a very high proportion of the incident radiation is reflected, scattered or absorbed outside the area covered by the mull, and the overall energy reaching the detector is correspondingly lower than anticipated. The effect on the spectrum is an apparent high background absorption, and it can be difficult to distinguish faults

1 and 3 without some knowledge of the sample. The main distinguishing feature is obviously the intensity of the normally weak fine-structure.

Sample Change

There is no simple solution to the problem of sample change because it can always occur with an otherwise well-behaved sample giving spectra of excellent quality. Real difficulties occur if it is superimposed on a technical fault, solution from first principles being the only satisfactory answer.

1. *Polymorphism.* Polymorphism is much too common to be dismissed lightly. It must always be anticipated if the sample has undergone *any* process different from normal. Fortunately, there are usually certain diagnostic bands common to all polymorphs and these may be recognized with experience. Careful scrutiny of all the spectra indicated by the index should reveal these diagnostic bands, even though the polymorph is not recorded.

Although not a very satisfactory solution—it relies as much on intuition and hunches as on science—study of the diagnostic bands will return an answer when a nil score is otherwise a certainty. A study of these patterns is invaluable when mixtures and impurities are encountered.

2. *Ion exchange.* Some laboratories only use pressed disks for sample presentation and must always expect ion exchange to occur when there is no common ion between sample and matrix. Sulphates and organic acid salts are the more obvious culprits to watch.

Changes in moisture content can promote or hinder exchange, and by its very nature, the effect on ion exchange can be very variable and unpredictable. Most of the spectral changes are obvious, and due allowance made for them in feeding information into the retrieval scheme, when it should be found that the success rate is vastly improved.

In a correctly maintained index, all the common examples of ion exchange will have been incorporated as they are encountered in every-day working. It is more of a problem when dealing with a complete unknown and the more logical matrix is not chosen straight away; in the course of time it will become less of a problem.

3. *Change of state.* The heat from incident radiation can melt samples, particularly in pressed disks subjected to beam-condensed radiation. Most reference spectra are recorded under the best conditions possible and there may not be a specimen taken under these more extreme conditions. Some of the amine salts will melt at quite low temperatures, particularly if in admixture with a small amount of impurity, and sample melting must always be anticipated when an extract from an unusual original sample is being studied by beam-condensing.

Spectra from molten material are generally simpler than from crystalline solids and can be recognized. Changes in wavelength of some otherwise diagnostic bands are not simple to correct for indexing.

The reverse process is not unlikely and may be regarded as a special variety of polymorphism. Some materials are converted to an amorphic form during preparation and a spectrum recorded for this physical state. Excessive irradiation could induce a reconversion back to a crystalline form which may not be recorded in the index, or, alternatively, the conversion could proceed spontaneously and be time dependent. Changes of state can occur in mulls if these are left for an appreciable time, say, over lunch or over night, regardless of whether or not they are left in an energy beam. Such changes are best recognized by an unexpectedly good resolution with relatively few extra bands, too few to be explained as gross impurity, and the relative weakness of some of the normal bands, to the extent, perhaps, of their total absence.

Combinations of all these nine causes are possible and are more common than they ought to be. A really critical appraisal of spectral quality is an essential adjunct to any retrieval system, and in the course of time, other common faults will be recognized by the user and added to these nine common ones.

Even with limited indexed data, the success rate can be dramatically improved by assessing spectral faults and quality. A limiting factor will always remain the accuracy of indexing; spot checks can be revealing.

Interference

There is little to be said about this type of correction or allowance.

It is not at all easy to correct for interference once a sample cell has been dismantled, and next to impossible from published spectra without intimate details of the sample extraction/preparation and the conditions prevailing during the spectral run. In the absence of this vital information, one is very much in the dark and can only guess at the basic cause of an unfamiliar spectrum, which could be due to a new physical form of a familiar item, an impure form of the same familiar item, or simply a totally different material.

Over a given period of time, it is inevitable that a number of the more ubiquitous impurities will be encountered and, it is hoped, noted, and the worker will gain experience of the type of correction to apply. Direct superimposition of a typical interference spectrum in a light-box is an extremely convenient way of assessing the degree of band distortion. For a simple qualitative identification, this is adequate (although difficult to convince a non-spectroscopist on occasions), but quantitative measurements are not simple. The basic rules of quantitative adjustment are well documented and a technique for estimating or establishing the true base-line in a similar connection is relevant here.[46]

It will take some time before all infra-red data are assessed objectively. Although not strictly the point being made here, some moves are being made through the Joint Committee on Atomic and Molecular Physical Data towards this end, where their primary concern has been an evaluation of available data in terms of classes of certainty.[6,7] Approximately half of the thousand substances examined are pharmaceuticals of class III, that is, compounds with a verified structure. A similar grading of spectra is long overdue.

For those wishing to delve into the theoretical aspects of band shapes and intensities and learn more about factors not discussed in these practical notes, the review of Seshadri and Jones is recommended.[175]

Checking the Index

Reference has been made elsewhere (p. 175 ff) to the necessity of checking, and double checking, at all stages. In concluding this sub-section on corrections, it is pertinent to repeat and emphasize the point.

One of the more difficult aspects of data appraisal is a decision on the grade of staff to employ in the collection and clerking of raw material, in this case spectra. In a small laboratory, the problem does not arise because only one or two persons are free to do all aspects of the work, and relatively highly qualified staff will be employed in clerking exercises, whilst at the other end of the scale, clerks will be elevated to doing scientific work. Some compromise is obviously needed for an entirely satisfactory answer.

Since a double check is mandatory, it is suggested that all the tedious sifting, clerking and punching be done by a competent junior member of the staff (preferably non-scientific), and that, whenever possible, should be given batches of spectra, not single ones at irregular intervals. The same batch of spectra is then given to one of the people actually using the spectrometer, who is then told to retrieve each spectrum in turn from the index prepared by the first person, whether this is a manuscript log-book, or a sophisticated computer miles away on the end of a telephone line. Provided the initial instructions are simple and clear, 100 per cent 'hits' should be recorded at this stage, and if so, do not rest on your laurels and neglect to do this chore each time a new batch is processed, because, sooner or later, a 'miss' is inevitable.

If one is to get any value out of the check, each 'miss' must not be dismissed as 'just one of those things', but thoroughly investigated by discussion between the persons concerned. The majority of 'misses' will be due to simple transcription errors, but even these should not be ignored—who is more prone to error, the initial sifter or the final operator? or can the punching regimen be simplified to eliminate spurious answers? or do the instructions issued give too much latitude for subjective errors? or is any other factor operative?

Once the transcription errors have been eliminated, the much more interesting cases of different data interpretation can be discussed. Again, which interpretation is more valid, or are both valid on the basis of the instructions? If both are valid, can the instructions be made more specific, or does a double or multiple entry need to be made in the index?

Much more important is a consideration of what caused the spectrum to be open to multiple interpretation. Is it a result of faulty presentation? or has the operator subconsciously utilized superior knowledge to correct it? or has something else subjective crept into the assessment? Consult all other colleagues until all parties are agreed, or agree to differ. In any event, the operator will learn much through discussion of the failings of the system and his own failings in preparing the spectrum, factors which will be slow to materialize if all indexing is carried out solely by himself or entirely devolved without checking. The quickest way of learning something is to teach it or to prove to a lay-person (the clerk) why he has made a mistake.

The value of employing persons not directly concerned with the ultimate application of the index cannot be overemphasized, because, to be of any real merit, an index must be immediately effective in the future by one's successors, who will have differing prejudices and subjective assessments from one's own.

The argument that not all laboratories will take the trouble to carry out checks of the type discussed here is indefensible.

The symbiotic relationship between two divergent approaches to the subject is of immense gain to both, in particular, in confirming that the final, agreed index is indeed 'idiot proofed'.

BAND ASSIGNMENT

Correlation charts for band assignment abound in the literature and so this section will be restricted to general remarks.

If the spectrum of an unknown is identified from the retrieval scheme as that of a single compound or a known mixture, further confirmation is not needed. This presupposes that all criteria are satisfied with respect to band position and relative intensities, in short, that the 'pattern' is correct. The presence of the occasional 'foreign' band is normally ignored or ascribed to a modified crystal structure of the material. It is in situations where these 'foreign' bands must be identified, or where the retrieval system fails, that band assignment comes into its own. The experienced worker will, of course, be capable of applying the band assignment approach first in preference to the retrieval system.

Major bands of the unknown are ascribed by referring to one of the published charts. Bellamy's books[14] are extremely comprehensive and cover most eventualities since nearly all materials studied in research institutes are not encountered in routine analysis. One would not expect, for instance, to encounter non-commercial substances like the exotic organometallics, although they are a fertile subject in the literature.

Books are not a very convenient reference source at a bench and one of the 'potted' charts is a frequent choice. Colthup charts[43] seem almost universal and appear as commercially produced, wall-mounted items, or in a variety of home-made guises, one being a simple photocopy (not always mounted on card) produced locally as needs demand. Copies reproduced on transparent sheets have been described, and in addition to being very durable, they can save a considerable amount of subjective assessment if reduced to the same width as one's own charts—simple sliding up and down will indicate immediately whether bands are in the correct position without calculation.

A more compact chart was described by Palmer[150] and is now available commercially.[93] This is sufficiently detailed to cater for all normal requirements, without going into the finer points and confusing the issue.

These simple correlation charts by Colthup and Palmer are calibrated linearly in wavenumber and wavelength respectively, and some mental dexterity is needed to inter-convert scales where these differ from the spectrometer charts. Tables of reciprocals are printed in most textbooks, whilst a good slide rule is usually to hand and more convenient than the conventional book of four-figure tables. In situations where the Sadtler notation is used for band assignment, or simply as a predigested inter-conversion to linear wavelength scales, the table published by Kramer will prove invaluable.[103]

Wavelength Check

Band assignment depends on the band position being measured accurately. The same is true for data retrieval. With preprinted charts and limited controls on a modern spectrometer, this measurement ought to be relatively easy, and proves to be the case in most instances. Some systematic errors are always possible, particularly over an extended period, and it is a wise precaution to check the wavelength settings at regular intervals.

The preprinted charts are (at least in theory) checked for accurate setting relative to the main spectrometer drive each time they are used, by reference to a fixed fiducial mark; errors of non-linear wavelength drive are rarely checked throughout the whole range. It has become the fashion to supply with each spectrometer a thin film of polystyrene, for the wavelength check, and this is often mounted in a large mask with a list of the major absorption bands printed on it. Provided this object does not get lost, there is no necessity for anything more elaborate.

Comprehensive instructions and data for a complete calibration with respect to indene are given by Jones and his co-workers.[97] This is a good method of familiarizing oneself with the capabilities and limitations of a spectrometer, and is an essential requisite if bands are measured to the nearest wavenumber as in Clarke's retrieval scheme.[38] Errors of a single wavenumber in a non-linear manner across the chart will invalidate the whole scheme; errors of considerably more than this are the rule rather than the exception and an accurate calibration chart or table must be drawn up and checked regularly. Micro-charts are just impossible for this level of accuracy—the width of the pen is outside the tolerances! Wavelength settings calibrated on a large chart do not necessarily apply on any other unless this is confirmed.

Chemical Approach

No deductions of chemical structure are of any use unless due consideration is taken of other chemical evidence. Useful clues to identify are given by the isolation method, which is designed to perform this function, and if this is based on some chromatographic separation, one or two strong possibilities will be indicated. Identification of a base from an extraction designed to isolate acids, or *vice versa*, is plainly ridiculous and yet is not unknown from experienced analysts. All factors must be taken into consideration, with particular regard to the circumstances surrounding the reasons why a sample is submitted for analysis.

Over twenty years ago, Barnes and his colleagues[12] reviewed the chemical separation approach, and their conclusions remain as true as ever. Alha and Tamminen[1] are particularly clear in their exposition, and this paper should be read by all.

Other workers have followed this line to a variable extent. Hofmann and Ellis[90] take it one stage further and make a salt derivative of the unknown *in situ*, the extra salt bands providing a useful further confirmation.

Every rule has its exceptions and, in the case of the chemical approach, amphoteric or neutral compounds may creep into the (apparently) incorrect chemical class. These are recognized very quickly if proper research is undertaken with pure materials before real samples are handled.

Even in this day and age, the human brain remains the most versatile data retrieval system known, by virtue of its ability to match patterns and weigh divergent information critically. Many characteristic patterns are recognizable 'on sight' in a manner quite impossible to programme for a computer, each person having a different, unconscious method of tackling the problem built up over a period of time.

In short, there is no substitute for experience in making a correct band assignment.

'KNOW THINE ENEMY'

One of the unkind apocryphal tales circulating about infra-red spectroscopy is that '. . . provided one tells an infra-red worker the answer, he will tell you the answer . . .'. There is an element of truth in this, but no more so than for any other analytical method. Until experience is gained with a particular material or group of materials, spectra can only be predicted on the basis of information already to hand and must be confirmed by direct experimentation. In this respect it is the same as a chromatographic run or a colour test. Once the basic information has been revealed, it can be filed away for future reference or published.

The simplest abstract or reference to the literature is provided by the annual reviews of *Analytical Chemistry*. Details of infra-red techniques are given in even-date years, and the more specialist applications can be sifted in odd-date years.

However, spectral information is not always culled from the literature, and, to make matters worse, authentic reference sample specimens may be unobtainable in the time allowed for an analysis, or, in the ultimate, supply restrictions in another country render it impossible to obtain at all. Some prior knowledge of the physical and commercial properties of all substances of potential forensic interest is therefore of paramount importance. Valuable clues to substantiate information derived from first principles can be gleaned from a number of sources, the following being, necessarily, only a selection. The selection is of books suitable for forensic work and those readers to whom this selection is of no interest should turn to p. 204.

Pure Materials

1. *Merck Index*. This covers all materials of forensic interest and a host of others of lesser concern.[133] The cross index and data on physical constants make this the most valuable single book. The number of trade names listed after the primary name and the nature of the primary name are a very good guide to the likelihood of the substance appearing on the market and being available in the form encountered. The *Merck Index* is up-dated at regular intervals and if a substance is not listed, it is a safe conclusion that it is of recent development, in which case it will be of topical interest and some information will be to hand from some other topical source.

2. *Martindale*. Although of more restricted coverage than the

Merck Index, Martindale's *Extra Pharmacopoeia* covers all the mundane drugs and a number of research and obsolete ones.[123] Its main value lies in the clinical data supplied with each major monograph, which includes details of toxicity, symptoms and typical case histories. This evidence can be crucial in selecting groups of compounds for study, as medical symptoms and case histories are often submitted in poisoning cases and samples from a post mortem.

3. *Poisons and TSA Guide.* The 'TSA Guide', as it is often misnamed, is irreplaceable for information on the legal status of pure drugs and the common medicaments made from them, data which are difficult to acquire elsewhere.[153] Cumulative amendments are published each month in the *Pharmaceutical Journal*, thus making it really up-to-date. By the very nature of the problem, the 'TSA Guide' cannot hope to cover all compounded commercial medicaments (very true of foreign products), but it does embrace all the major ones and is seldom found wanting in this respect. Reference to the *Merck Index* or 'Martindale' usually provides an answer for an obscure medicament by indicating major components which can then be studied in the 'Guide' for legal status. The restriction on the sale of a medicament is an indication, albeit a crude one, of the probability of its being found in the sample.

4. *Multilingual List of Narcotic Drugs under International Control.*[148] The three previous reference books are printed in English only and cover relatively few substances marketed outside Great Britain or America. This UN booklet is a unique collection of synonyms, in the appropriate language, of all the internationally controlled narcotics, and no forensic science laboratory should be without a copy. With an increasing international trade, not all of it licit, problems of translation are greatly simplified by this publication, an important point where the penalties for illicit possession are becoming increasingly severe.

A pleasant feature of the list is the reproduction of structural and molecular formulae.

5. *World requirements.* Statistics are published from time to time of the anticipated world requirements of narcotic drugs,

together with details of legitimate production. These figures on production and usage for licit purposes are a useful guide to the possibility of encountering some of the more exotic alkaloids; the more popular ones are obviously grossly in error in their figures because of illicit sources and uses, but these are already familiar through their popularity and there will be ample analytical data to hand.

The *Bulletin on Narcotics*, published by the UN always carries a comprehensive tabulation of the 'world requirements' at regular intervals, while other journals print abstracts.

Compositions

The identification of tablets and their active ingredients is simplified by considering their appearance. The 'ballistics' of all common tablets and capsules are tabulated in a number of books, and with an increasing tendency for firms to make their products distinctive, with or without code marks, unique identification is often possible before any chemical work is started. The larger laboratories will also possess comprehensive collections of solid-dose preparations, with their own methods of retrieval, and they can select an authentic specimen for direct comparison with the suspect.

Whether authentic specimens are to be used or not, the information given in the various catalogues is irreplaceable. Direct assays or identification tests can be applied without the tedium of a full and comprehensive analysis, and among these direct identification tests infra-red analysis is pre-eminent. Characteristic bands from one of the indicated ingredients may be identified in the morass from the excipients, in a direct examination, if a reference spectrum of the ingredient is superimposed on a light-box. Further confirmation can be found by extraction/isolation of this ingredient, colour tests, or a host of other methods. Another method, one of the simplest, is direct chloroform extraction and examination as a smear on a rock salt plate; it has the advantage that excipients do not, as a rule, interfere. This technique is described on p. 97 as a candidate for routine analysis, and on p. 179 as a candidate for an independent data retrieval system.

No catalogue of tablet/capsule ballistics is comprehensive and none attempt to cover more than the products of one country in any detail. Only three find wide usage.

1. *Chemist and Druggist's Tablet Identification Guide.* This guide indexes separately by name and colour code the majority of British tablets and capsules[31] and a regular amendment to it is published in the first issue each month of the *Chemist and Druggist.* It is remarkably comprehensive, only omitting 'fringe medicines' for the most part, and the products of the occasional uncooperative manufacturer.

The method of indexing chosen is by colour against a standard chart. Although it could be argued that the colours chosen are purely arbitrary, this is of no consequence when contending with colour blindness because an exact hue (or near match) can be assessed from the chart provided. The truly colour blind person will always be lost whatever is chosen for a reference chart, unless some instrumental aid is sought.

Non retrieval of an unknown tablet with a distinctive characteristic indicates, with a fair certainty before closer inspection, that it is of recent manufacture, a 'fringe medicine' and therefore relatively harmless, a confection, or of foreign manufacture.

2. *Physicians Desk Reference.* Published annually in the U.S.A. this book gives very detailed information on all the major pharmaceutical companies' products circulating in that country.[152] A section in colour illustrates a high proportion of the solid-dose preparations at full size with their respective names, and, in a number of cases, with a note of their major components. Several cross-indexes are listed together with a comprehensive reprint of medical and other data supplied by the companies contributing to the book.

No laboratory concerned with the analysis of tablets should be without a copy of 'PDR'.

3. *Medindex. Medindex*[131] does not attempt the coverage or cross-indexing of the American equivalent. However, it does contain sufficient original material to merit purchase.

These three books will not be the only ones chosen by most laboratories, but will be supplemented by a variety of trade journals and charts from individual firms, and most laboratories will also build up their own collection of authentic tablets. Provided a trade name or some commercial name is divinable from the facts surrounding any particular sample, details of the composition can also be found in a number of other publications.

4. *Monthly Index of Medical Specialities.* Known as 'MIMS',[146] this publication gives, in a very concise form, basic and essential information on all the normal proprietary materials prescribed by general practitioners. The cross-indexing for specific medical applications and the method of listing present all proprietaries of like type in groups for ease of reference. The circumstances in which a sample is found may lead one to make a detailed search through the relevant section of 'MIMS', and from this to a small group of probable medicaments.

5. *Chemist and Druggist Price List.* Regular weekly supplements, on a cumulative basis, are supplied for this quarterly publication.[30] Once acquired, this list is invaluable for assessing the current market status of a proprietary—in this case not necessarily a drug—and for identifying a manufacturer where his product is previously unknown. Having identified a manufacturer, he is invariably co-operative in dealing with *bona fide* enquiries and in supplying vital information difficult or impossible to find anywhere else.

There is an annual companion volume to the quarterly price lists[32] which is of more value in locating sources of pure drugs and summarizing the legal position on them.

6. *Chemical Toxicology of Commercial Products.* This is a massive tome[42] which is devoted to poisons and thereby gives some information on symptoms and toxicity in a convenient format. It has a few errors in the various sections on the market availability of the materials, but of no real consequence in this context; it has a distinct bias towards American products, which becomes of some advantage in Great Britain when imports are considered.

7. *Other sources.* Provided a trade name can be identified, the *Merck Index* and 'Martindale' are sufficiently detailed and comprehensive for nearly every eventuality, at least as far as the confirmation of major active ingredients is concerned.

The *Condensed Chemical Dictionary*[44] is too expensive and space consuming to be possessed by most laboratories, but within the limits imposed by its terms of reference, the information within its covers should give every satisfaction.

General Literature

A comprehensive coverage is a major undertaking in itself and beyond the capacity of most if the word 'comprehensive' is taken literally. The remarks by White on this topic[201] cannot be bettered and help to bring the affair into perspective.

Technical articles and specific applications appear at regular intervals in *Analytical Chemistry*, *Applied Spectroscopy*, and *Journal of Pharmaceutical Sciences*, and at very irregular intervals in *The Analyst*, *Journal of Pharmacy and Pharmacology* and many other journals; all important sources may be culled from *Analytical Abstracts*.

In concluding this section on information retrieval, it is to be hoped that some more collaborative effort will be devoted to data storage and retrieval, and the efforts of at least some of the individual researchers itemized here will be rewarded. The commercial items are to be commended (and supported), but until adequate notice is taken of the quality of data freely published, and held by others to be 'excellent', 'comprehensive' and other eulogistic phrases, the value of all attempts is seriously diminished.

Until the halcyon days dawn, the average worker is probably better off making his own arrangements based on his own assessment of the data, which, in turn, will be based on his own experience and competence.

FOOTNOTE ON INFORMATION RETRIEVAL

There is a marked danger in discussing information retrieval which is sufficiently common to merit this antidote.

The proverb 'They cannot see the wood for the trees' is as true of infra-red data retrieval as in any other topic. Some people are obsessed with systems as ends in themselves, and lose sight of the primary objective of data retrieval, namely, the acquisition of an authentic spectrum of the substance being examined for direct comparison. Until this comparison is made and all the essential criteria are satisfied, no system can be said to have been successful.

The mere presentation of a named compound from the index is not good enough, and the operator must satisfy himself that the spectrum relating to this named compound corresponds to the unknown 'beyond all reasonable doubt'. How one reaches this stage of comparing spectra is of little consequence, as long as the system is reliable and works to one's satisfaction. The person who relies on his brain, memory, knowledge of chemistry, and a reliable band assignment table (and, dare one say it, thinks first, speaks last), is as important a part of infra-red data retrieval as a highly sophisticated computer complex.

If anyone knows of a person obsessed with data retrieval, why not consider hanging this notice above the spectrometer?

THE OBJECTIVE OF DATA RETRIEVAL IN
INFRA-RED ANALYSIS IS

THE DIRECT COMPARISON OF TWO SPECTRA

NOT

BIGGER AND BETTER SYSTEMS

Our little systems have their day
they have their day and cease to be.
TENNYSON

QUESTIONS

Patience, stout heart, thou hast endured even
worse ills than this.

HOMER

A few questions have been chosen to amplify some of the more common spectral faults, and to test the efficacy of retrieval systems. To obtain the maximum benefit from these questions, be honest in your answers and do not turn to p. 211 until you have finished. Remember, you will not be able to turn to p. 211 when being cross-examined in court!

1. In all the pressed-disk spectra presented as examples, there is one major procedural fault which has not been mentioned. Identify this omission.

2. Code all the methylamphetamine spectra presented as examples, in any scheme you choose, and compare the results with your own spectra and those extracted from the literature.

3. Code the methadone spectra presented as examples, in any scheme you choose, and compare the results with your own spectra and those extracted from the literature.

4. Code the diamorphine hydrochloride spectra presented as examples, in any scheme you may choose, and compare the results with your own spectra and those extracted from the literature.

5. Code the phenmetrazine hydrochloride spectra presented as examples, in any scheme you may choose, and compare the results with your own spectra and those extracted from the literature.

6. Code the promethazine hydrochloride spectra presented as examples, in any scheme you choose, and compare the results with your own spectra and those extracted from the literature.

20. This substance is related to the previous three. With this knowledge, identify from first principles.

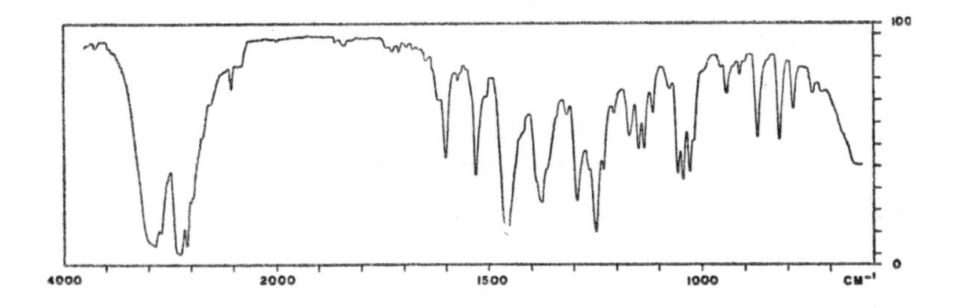

21. The previous 20 questions were concerned with exercises in information retrieval, or with the identification of procedural faults in examples specifically prepared by myself as questions. Before leaving the subject of procedural faults, it is necessary to select a 'real' example from the literature. Comment on the spectra presented by O'Brien, K. P., and Sullivan, R. C., 'Cocaine and its substitutes studied with infra-red spectrophotometer', *Bull. Narcot.*, 22(2), 35–40.

Cocaine was one of two drugs singled out for extra control by the Dangerous Drugs Act, 1967.

ANSWERS

1. No details of sample concentration are given. To obtain the maximum information from a pressed-disk spectrum, the sample concentration is an important parameter. The spectra taken for illustration are intended to show one facet of technique at a time.

2. The *AOAC* series reference number is 301.[211] This has the same relative band intensities as Clarke's spectrum, and the original spectrum used for Curry's index (Ref. No. 691), viz: 13·3, 14·3, 9·4, 6·2, 9·2, 9·6 μm.

Curry has indexed an extra band at 8·4 μm, and this will allow for some variation in the 9·2 to 9·7 μm region.

3. The *AOAC* series reference number is 220. This spectrum has 6 major bands at 5·8, 14·1, 13·0, 9·0, 8·8, 9·2 μm. Clarke does not give a spectrum.

Curry gives these bands: 14·1, 5·9, 13·0, 9·0, 10·6, 8·8 μm.

4. No spectrum of diamorphine hydrochloride is given in the *AOAC* series or by Clarke. There are many examples of both polymorphs in the literature, although not cited as polymorphs.

5. The *AOAC* spectrum No. 468, that of Clarke's on p. 754, and Curry's spectrum are recovered from the same 6 bands, but each in a different order of intensity.

6. All the readily available spectra of promethazine hydrochloride show a different relative intensity of the 6 major bands.

7. The same pellet used for Fig. 7.60 in the examples, run with a fast scan and normal slit settings. Compare with this a spectrum run with a slow scan and narrow slits. An instrument designed for routine work can compare favourably with a more sophisticated version if it is correctly 'tuned'.

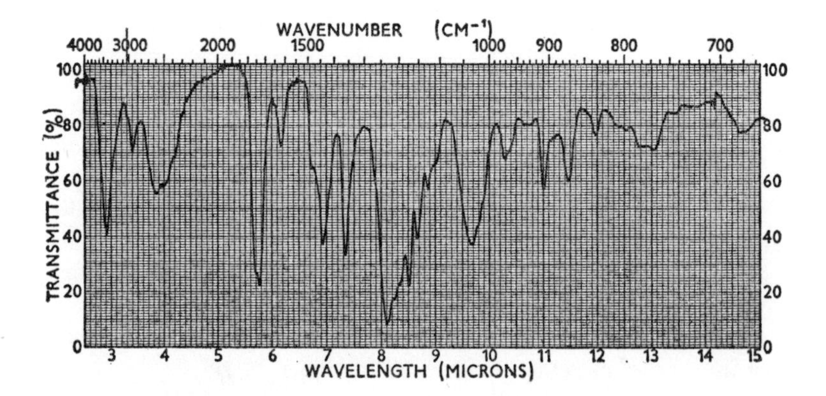

8. Dihydrocodeine hydrogen tartrate. A number of alkaloids of this type have a weak spectrum, and the 'cramped' spectrum is not due to incomplete beam-filling. Comparison with a mull spectrum confirms ion exchange:

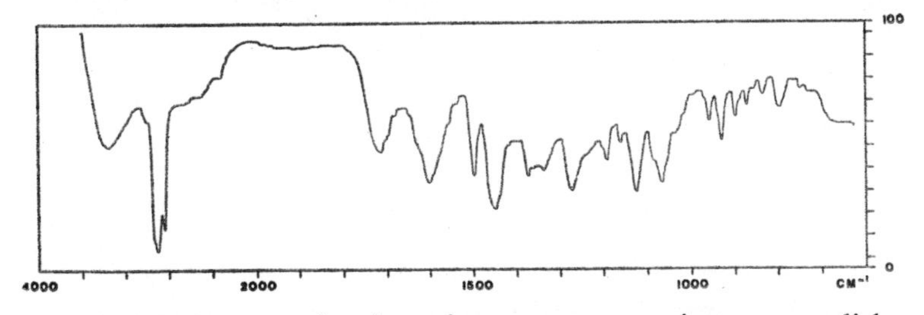

9. Ephedrine, as a free base, is very prone to give opaque disks, owing to an incompatibility with the common matrices. Its softness may contribute to the effect. The example here has not been well mixed. Better results are obtained with a paraffin oil mull:

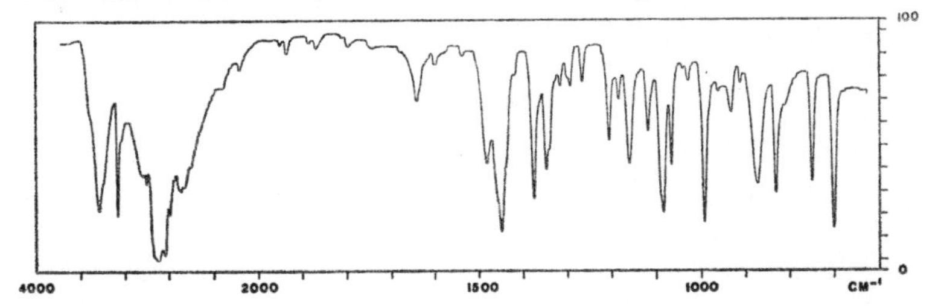

10. Pethidine hydrochloride presented as a smear on a NaCl disk, from chloroform solution. The substance does not crystallize quickly and a similar spectrum can be obtained in a pressed disk from a compound evaporated from chloroform solution. A mull spectrum is quite distinct:

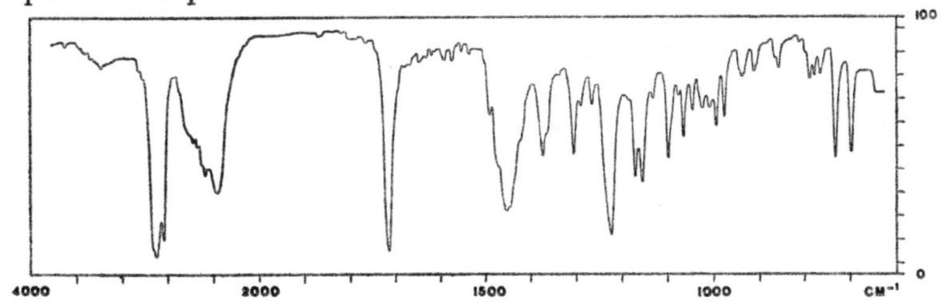

11. Poor grinding and/or dispersion. Relative band intensities are altered, and the band at 1245 is subject to Christiansen filter effects. The new 6 most intense bands are: 1650, 1248, 1293, 760, 1210, 1155 cm^{-1} (uncorrected, but on the previous chart settings; other bands are prone to frequency shifts):

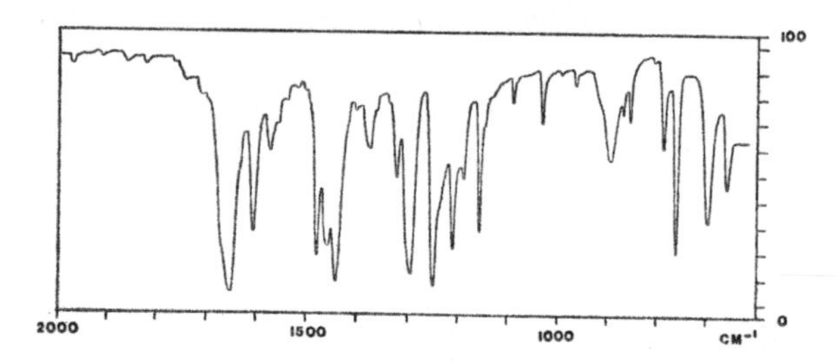

12. Cocaine hydrochloride. Poor mixing of a relatively well ground material. The chart paper has been displaced, i.e. *not* correctly set, by about 5 cm^{-1}. A mull spectrum is given here for comparison.

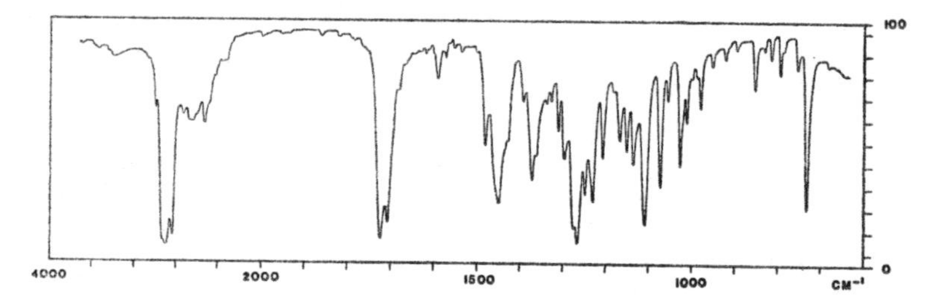

13. Diphenhydramine hydrochloride. Further grinding produces this spectrum.

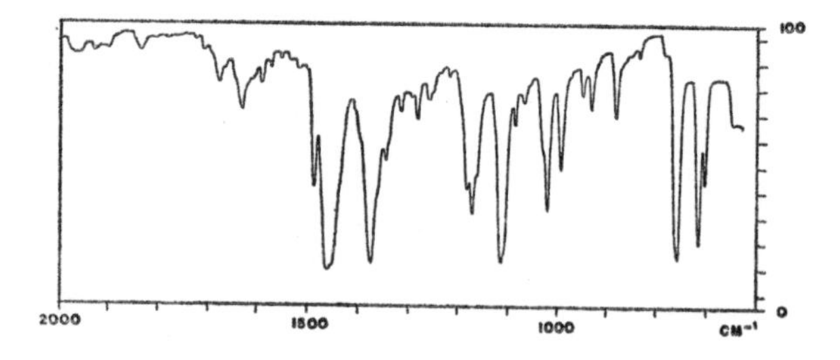

14. Methylphenobarbitone in a paraffin oil mull. Bands from the suspending medium are not always obtrusive.

15. Thiamine hydrochloride. A better mull:

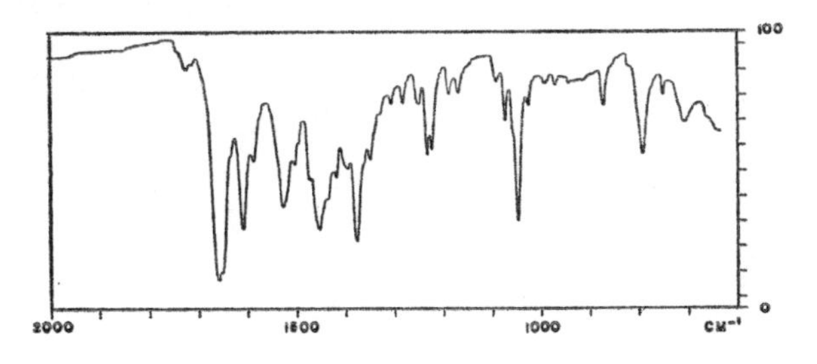

16. The differing spectra indicate polymorphism or variations from manufacturing processes. The spectrum here is of methaqualone from a major supplier in the U.S.A.:

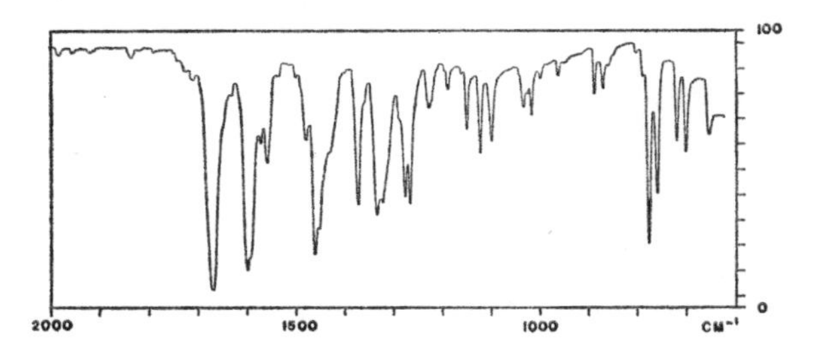

17. Ephedrine hydrochloride in paraffin oil mull.

18. Pseudephedrine hydrochloride in paraffin oil mull.

19. Phenylephrine hydrochloride in paraffin oil mull.

20. Isoprenaline hydrochloride in paraffin oil mull.

21. In my opinion the spectra show (*a*) excessive interference from the mixing vessel, (*b*) poor mixing (see Fig. 7.54, p. 148), and

(*c*) poor grinding. Some of the spectra show evidence for ion exchange with the matrix. A spectrum of butacaine sulphate in a paraffin mull is presented here for comparison:

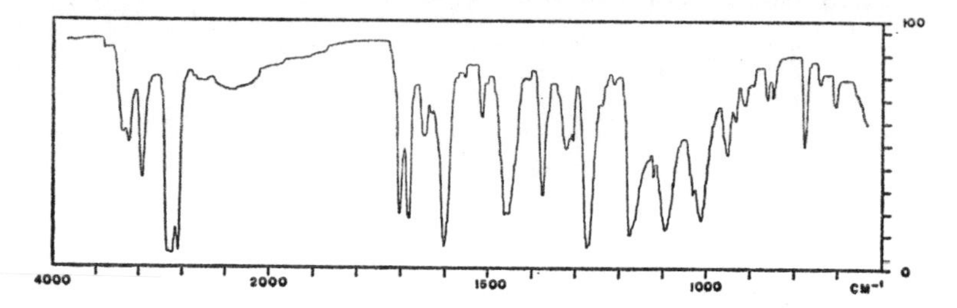

BIBLIOGRAPHY

When found make a note of.

DICKENS

The following articles appear to have some relevance to the theme of the book, although they have not been referred to in the text.

General

WILKS, P. A., Survey of IR Spectrophotometers, *Instrum. Contr. Syst.*, 1968, 41 (8), 67–71.

LUONGO, J. P., Continuous reference beam attenuator for IR spectrophotometry, *Appl. Spectros.*, 1960, **14**, 24–25.

LIPPENCOTT, E. R., WALSH, F. E. and WEIR, C. E., Micro techniques for the study of solids; diamond and sapphires as cell materials, *Analyt. Chem.*, 1961, **33**, 137.

KIRKLAND, J. J., IR spectrophotometric analysis of fractional microgram quantities of solids, *Analyt. Chem.*, 1957, **29**, 1127.

LOW, M. J. D., *et al.*, Characterization of organic compounds by means of their self-emitted IR radiation, *Chem. Commun.*, 1965, (16), 389–390.

GOULD, J. H. and CHEN, J. T., A nomograph for the correction of spectral data recorded in transmittance, *Appl. Spectros.*, 1969, **23**, 550–551.

COOKE, P. A., Unicam IR spectrum interpretation aid, *Spectrovision* (Pye Unicam Ltd, Cambridge), 1970, No. 24, 2–3.

Technique

SANTOS, A. S., Practical aspects of IR, *Rev. Port. Farm.*, 1965, **15**, 424–443 (Port.).

TOLK, A. and VAN DER MOLEN, H. J., The influence of the preparation technique upon the IR spectrum of benzoic acid, *Chem. Weekblad*, 1957, **53**, 656–657.

JENSEN, J. B., Determination of solids by means of IR spectroscopy and potassium bromide disk technique, *Dansk. Tidsskr. Farm.*, 1958, **32**, 205–220; 221–235.

JENSEN, J. B., Anomalies in IR spectra using the KBr disk technique, *Dansk. Tidsskr. Farm.*, 1959, **33**, 33–34.

ROPKE, H. and NEUDERT, W., Advantages and limitations of the KBr method in IR spectroscopy, *Z. Anal. Chem.*, 1959, **170**, 78–95.

SCHIELE, C., Deviation of band contours by various halides in the pellet technique, *Appl. Spectros.*, 1966, **20**, 253–258.

BERGSTEDT, E. I. M. and WIDMARK, G., The KBr 'eye' technique; How to analyse by IR one microgram of sample in a KBr pellet pressed in an ordinary 13 mm die, *Chromatographia*, 1970, **3**, 216–219.

COPIER, H. and SCHUTTE, L., Efficient trapping and transfer of microgram gas chromatographic fractions for IR analysis, *J. Chromatog.*, 1970, **47**, 464–469.

TSUDA, T., and ISHII, D., Combination of a gas chromatograph and a fraction collector using a small condenser, *J. Chromatog.*, 1970, **47**, 469–473.

MATOS, G., Application of IR spectroscopy to gas chromatography. I, Problems appearing in sampling at the exit of the chromatographic apparatus with silver chloride cell, *Afinidad*, 1970, **27** (275), 171–183.

Information Retrieval

KUENZTZEL, L. E., New codes for Hollerith-type punched cards. To sort IR absorption and chemical structure data, *Analyt. Chem.*, 1951, **23**, 1413–1418.

ERLEY, D. S., Card coding of spectroscopic data, *Appl. Spectros.*, 1969, **23**, 548–549.

General Applications

PRICE, W. C., IR spectroscopy and its applications to pharmaceutical analysis, *J. Pharm. Pharmac.*, 1955, **7**, 153–166. A review with correlation charts.

BOUCHE, R., Drug analysis by IR spectrophotometry, *J. Pharm. Belg.*, 1968, **23** (9–10), 449–502 (Fr.). Review with 370 references.

Specific Applications

Nearly all the references cited in this section include spectra. The exceptions include tabular material or have only an infra-red slant.

Dangerous Drugs Act materials and close relatives

PLEAT, G. B., HARLEY, J. H. and WIBERLEY, S. E., Use of IR spectra in the qualitative and quantitative determination of alkaloids, *J. Am. Med. Ass., Sci. Ed.*, 1951, **40**, 107–110.

MANNING, J. J., Pressed bromide method of IR spectrographic analysis of narcotics, *Appl. Spectros.*, 1956, **10**, 85–98.

LATSHAW, W. E. and MACDONNELL, D. R., Phenazocine HBr: pharmaceutical aspects of a new benzomorphan analgesic related to morphine, *J. Pharm. Sci.*, 1961, **50**, 792–797.

GENEST, K. and FARMILO, C. G., Simultaneous determination of morphine, codeine and porphyroxine in opium by IR and visible spectrometry, *Analyt. Chem.*, 1962, **34**, 1464–1468.

LERNER, M. and MILLS, A., Some modern aspects of heroin analysis, *Bull. Narcot.*, 1963, **15**, 37–42.

BROCHMANN-HANSSEN, E. and NIELSEN, B., 6-methylcodeine—a new opium alkaloid, *J. Pharm. Sci.*, 1965, **54**, 1393.

LEFEBVRE, J. Ch., Applications of IR to the identification of active principles, *Labo Pharma*, 1966, **144**, 56–65 (Fr.).

SMITH, L. L., Molecular asymmetry of methadon, *J. Pharm. Sci.*, 1966, **55**, 101–103.

BECKETT, A. H., *et al.*, The biotransformation of methadone in man; Synthesis and identification of a major metabolite, *J. Pharm. Pharmacol.*, 1968, **20**, 754–762.

Other restricted drugs and close relatives

KORNFIELD, E. C., *et al.*, The total synthesis of lysergic acid, *J.A.C.S.*, 1956, **78,** 3087–3114.

GOUTAREL, M., *et al.*, Über das ibolutein, *Helv. Chim. Acta*, 1956, **39,** 742–748.

TROXLER, F. and HOFMANN, A., Substitutionem am ringsystem der lysergsäure: I—substitutionem am indole-stickstoffe; II—alkylierung, *Helv. Chim. Acta*, 1957, **40,** 1706–1720; 1721–1732.

STAUFFACHER, D. and SEEBICK, E., Voacangarin, ein neues alkaloid aus Voacanga africana stopf, *Helv. Chim. Acta*, 1958, **41,** 169–180.

HOFMANN, A., *et al.*, Psilocybin and psilocin, *Helv. Chim. Acta*, 1959, **42,** 1557–1572.

DUVALL, R. N., KOSHY, K. Y. and PYLES, J. W., Comparison of reactivity of amphetamine, methylamphetamine and dimethamphetamine with lactose and related compounds, *J. Pharm. Sci.*, 1965, **54,** 607–611.

CHOULIS, N. H., Intermolecular association of stereoisomers as examined by IR spectra, *J. Pharm. Sci.*, 1965, **54,** 1367–1370.

SALIM, E. F. and MARTIN, A. E., Qualitative and quantitative tests for doxapram hydrochloride, *J. Pharm. Sci.*, 1967, **56,** 748–749.

SALIM, E. F., *et al.*, Qualitative and quantitative tests for chlorphentermine hydrochloride, *J. Pharm. Sci.*, 1967, **56,** 1001–1003.

Barbiturates

MANSON, J. M. and CLOUTIER, J. A. R., Physico-chemical characterisation of some 5,5-disubstituted-1-3-dimethylbarbituric acids, *Appl. Spectros.*, 1961, **15,** 77–79.

PAULING, G., *et al.*, Anomalies in the IR spectrometric detection of barbiturates, *Z. Anal. Chem.*, 1963, **193,** 28–33.

WOLFSON, B. B. and BANKER, G. S., Solubilisation and evaluation of poly-N-vinyl-5-methyl-oxazolidinone barbiturate systems, *J. Pharm. Sci.*, 1965, **54,** 195–202.

FRENCH, W. N. and MORRISON, J. C., Identification of complexes of phenobarbitone with quinine, quinidine or hydroquinidine in pharmaceutical dosage forms, *J. Pharm. Sci.*, 1965, **54,** 1133–1136.

Other drugs

FRITZ, H., Uber die konfiguration am C–19 des serpentins, *Annalen*, 1962, **655,** 148–167.

RAPSON, H. D. C., AUSTIN, K. W. and CUTMORE, E. A., Analysis of poldine methyl methosulphate by IR spectroscopy, *J. Pharm. Pharmac.*, 1962, **14,** 66T–72T.

ELLIOTT, T. H. and NATARAJAN, P. N., IR studies of hydantoin and derivatives, *J. Pharm. Pharmac.*, 1967, **19,** 209–216.

Other examples of the application of infra-red to the analysis of drugs have been limited to a selection from the *Journal of Pharmaceutical Sciences* (and its precursor). They are given in date order:

WASHBURN, W. H., IR analysis of active ingredients in ointments; I, atropine sulphate and scopolamine HBr, 1952, **41,** 602–605.

ibid., II, pilocarpine HCl and phenacaine HCl, 1953, **42,** 698–699.

HALGREN, P. F., THEIVAGT, J. G. and CAMPBELL, D. J., Methods for the determination of tral, a new anticholinergic drug, 1957, **46**, 639–643.

HAYDEN, A. L. and SAMMUL, O. R., IR analysis of pharmaceuticals; II, A study of cinchona alkaloids in KBr disks, 1960, **49**, 497–502.

FELDMAN, E. G., *et al.*, Qualitative and quantitative tests for isobucaine HCl, 1961, **50**, 347–350. Correction and spectrum on p. 890.

AUERBACH, M. E., *et al.*, Qualitative and quantitative tests for mepivacaine HCl, 1962, **51**, 491–493.

WELSH, L. H. and SUMMA, A. F., Determination and identification of *p*-hydroxyamphetamine as the *O,N*-diacetyl derivative, 1963, **52**, 656–658.

ANON., Qualitative and quantitative tests for carphenazine maleate, 1964, **53**, 101–103.

ANON., Qualitative and quantitative tests for anisotropine methylbromide, 1964, **53**, 205–207.

TROUP, A. E. and MITCHNER, H., Degradation of phenylephrine HCl in tablet formulations containing aspirin, 1964, **53**, 375–379.

SMITH, E., Assay of glutethimide tablets, 1964, **53**, 942–944.

ANON., Qualitative and quantitative tests for chloral betaine, 1964, **53**, 1385–1386.

ANON., Qualitative and quantitative tests for metaxalone, 1964, **53**, 1522–1523.

ANON., Qualitative and quantitative tests for pipezethate HCl, 1966, **54**, 1338–1341.

BROCHMANN-HANSSEN, E., NEILSEN, B. and UTZINGER, G. E., Opium alkaloids; II, Isolation and characterization of codamine, 1965, **54**, 1531–1532.

DOAN, L. A. and CHATTEN, L. G., Identification and differentiation of organic medicinal agents II, 1965, **54**, 1605–1609.

ANON., Qualitative and quantitative tests for pargyline HCl, 1966, **55**, 618–619.

EISDORFER, I. B., Identification of 1-, 2-, 3-, 4-chlorophenothiazines isomers 1966, **55**, 734–735.

SALIM, E. F., *et al.*, Qualitative and quantitative tests for tybamate, 1966, **55**, 1439–1441.

SALIM, E. F. and ÖRTENBLAD, Qualitative and quantitative tests for prilocaine HCl, 1967, **56**, 1645–1646.

REFERENCES CITED

(Numbers in square brackets refer to pages where the references are cited.)

1. ALHA, A. and TAMMINEN, V., IR absorption spectroscopy in forensic toxicological practice, published in *Methods of Forensic Science*, Vol. IV, A. S. CURRY (ed.) (Interscience, 1965), pp. 265–298. [198]
2. ANDERSON, D. H. and COVERT, G. L., Computer search system for retrieval of IR data, *Analyt. Chem.*, 1967, **39**, 1288–1293. [165, 174]
3. ANON., Documentation of molecular spectroscopy (DMS system), *Appl. Spectros.*, 1956, **10**, 104–105. [164]
4. ANON., A card sorting/punching machine, *O & M Bulletin*, 1960, **15**, 125–126. *N.B.* A photograph is reproduced on the centre pages of the same issue. [181]
5. ANON., Absorption spectroscopy. IR spectroscopy, *Microchemical and Instrumental Analysis* (Millipore Corporation, 1963). [77]
6. ANON., The NBS standard reference materials programme, *Analyt. Chem.*, 1966, **38** (8), 27A–40A. [193]
7. ANON., Editorial comment, *Analyt. Chem.*, 1969, **41** (4), 97A–98A. [193]
8. Anson 'Visipoint' edge-punched card system (George Anson & Co. Ltd, Solway House, Southwark Street, London, SE1). See also Ref. 126. [180]
9. BAKER, A. W., Solid-state anomalies in IR spectroscopy, *J. Phys. Chem.*, 1957, **61**, 450–458. [69]
10. BARKER, S. A., *et al.*, IR spectra of carbohydrates; VI, Avoidance of spectral changes with KBr films, *Chemy. Ind.*, 1956, **16**, 318. [22]
11. BARNES, R. B., *et al.*, IR analysis of crystalline penicillins, *Analyt. Chem.*, 1947, **19**, 620–627. [52]
12. BARNES, R. B., *et al.*, Qualitative organic analysis and IR spectrometry, *Analyt. Chem.*, 1948, **20**, 402–410. [198]
13. BAUMAN, L. A., Jr., A technique for the use of the 'Mini-Press' in micro IR spectroscopy, *Appl. Spectros.*, 1969, **23**, 282–283. [95]
14. BELLAMY, L. J., *Advances in IR Group Frequencies* (Methuen & Co. Ltd, London, 1968). [196]
15. BELLANATO, J., IR spectra of ethylenediamine dihydrochloride and other amine hydrochlorides in alkali halide disks, *Spectrochim. Acta*, 1960, **16**, 1344–1357. [71]
16. BLACK, E. D., Potassium bromide capillary cell for IR microspectroscopy, *Analyt. Chem.*, 1960, **32**, 735. [50]
17. BLAKE, B. H., ERLEY, D. S. and BEMAN, D. S., Sampling technique for obtaining IR spectra of gas chromatographic fractions, *Appl. Spectros.*, 1964, **18**, 114–116. [76]
18. BLAKE, M. I. and SIEGAL, F. P., Analysis of phenobarbital elixir by an ion exchange and nonaqueous titration procedure, *J. Pharm. Sci.*, 1962, **51**, 494–496. [87]
19. BLOUT, E. R., *et al.*, IR microspectroscopy: III, A capillary cell for liquids, *J. Opt. Soc. Am.*, 1952, **42**, 966–968. [51]
20. BRADLEY, K. B. and POTTS, W. J., Jr., The internally standardized nujol mull as a method of quantitative IR spectroscopy, *Appl. Spectros.*, 1958, **12**, 77–80. [52]

21. BROWNING, R. S., WIBERLEY, S. E. and NACHOD, F. C., Application of IR spectrophotometry to quantitative analysis in the solid phase, *Analyt. Chem.*, 1955, **27**, 7–11. [54]
22. BUTZ, W. H., *Beckman Instr. Application Data Sheet*, 1960, IR–8063. [6]
23. CADMAN, W. J., IR examination of micro samples—application of a specular reflectance system, *Appl. Spectros.*, 1965, **19**, 130–135. [35]
24. CAMPBELL, J. A. and SLATER, J. G., Modification of physical properties of certain antitussive and antihistaminic agents by formation of N-cyclohexylsulphamate salts, *J. Pharm. Sci.*, 1962, **51**, 931–934. [91]
25. CHANG, S. S., *et al.*, A capillary trap for the collection of gas chromatographic fractions for IR spectrophotometry, *Appl. Spectros.*, 1962, **16**, 106–107. [76]
26. CHATTEN, L. G. and LEVI, L., An IR study of the reaction of barbiturates with *p*-nitrobenzyl chloride, *Appl. Spectros.*, 1957, **11**, 177–188. [92, 107]
27. CHATTEN, L. G. and LEVI, L., Identification and differentiation of sympathomimetic amines, *Analyt. Chem.*, 1959, **31**, 1581–1586. [90, 91]
28. CHATTEN, L. G., PERNAROWSKI, M. and LEVI, L., The identification and determination of some official local anaesthetics as tetraphenylborates, *J. Am. Pharm. Assoc., Sci. Ed.*, 1959, **48**, 276–283. [90]
29. CHATTEN, L. G. and DOAN, L. A., Identification and differentiation of organic medicinal agents; III, Amine containing antiParkinson agents and newer muscle relaxants, *J. Pharm. Sci.*, 1966, **55**, 372–376. [90, 91]
30. *The Chemist and Druggist Quarterly Price List* (Morgan-Grampian (Publishers) Ltd, London). [203]
31. *The Chemist and Druggist Tablet and Capsule Identification Guide: Mark 2* (*Chemist and Druggist*, Morgan Brothers (Publishers) Ltd, London, 1966). [202]
32. *Chemist and Druggist Year Book and Buyers Guide*, 1971 (Morgan-Grampian (Publishers) Ltd, London). [203]
33. CHEN, J. T. and GOULD, J. H., Micro AgCl technique for IR spectra, *Appl. Spectros.*, 1968, **22**, 5–7. [7, 49]
34. CHOUTEAU, J., DAVIDOVICS, G. and DEFRETIN, J. P., Spectrophotometric study in the IR of some long-acting sulphonamides, *Ann. Pharm. Franc.*, 1963, **21**, 487–499. [107]
35. CIC cavity cell Type C, *U.S. Patent* 3 036 215, May 22, 1962 (available from Barnes Engineering Co., Stamford, Conn., USA). [50]
36. CLARK, C., Cataloguing of IR spectra, *Science*, 1950, **111**, 632–633. [165]
37. CLARK, R. E., KBr disks for liquid samples, *Appl. Spectros.*, 1960, **14**, 139. [50]
38. CLARKE, E. G. C. and BERLE, J., *Isolation and Identification of Drugs in Body Fluids and Post-Mortem Material* (an *Extra Pharmacopeia* companion volume, general ed. R. G. TODD, The Pharmaceutical Press, London, 1969). [74, 165, 166, 197, 209, 211ff]
39. CLAUSON-KAAS, N., *et al.*, Note on the KBr disk technique for measurements of IR spectra, *Acta Chem. Scand.*, 1954, **8**, 1088–1089. [13]
40. CLEVERLEY, B., Identification of barbiturates from their IR spectra, *Analyst, Lond.*, 1960, **85**, 582–587. [70, 107]
41. CLEVERLEY, B. and WILLIAMS, P. P., Polymorphism in substituted barbituric acids, *Tetrahed.*, 1959, **7**, 277–288. [91, 107]

42. Gleason, M. N., *et al.*, *Chemical Toxicology of Commercial Products, Acute Poisoning* (The Williams and Wilkins Co., Baltimore, USA, 3rd edn, 1969). [203]
43. Colthup, N. B., Spectra-structure correlations in the IR region, *J. Opt. Soc. Am.*, 1950, **40**, 397–400. [196]
44. Rose, A. and Rose, E. (eds.), *The Condensed Chemical Dictionary* (Reinhold Publishing Corporation, New York, 7th edn, 1966). [204]
45. Copier, H. and van der Maas, J. H., Micro IR spectrometry of gas chromatographic fractions, *Spectrochim. Acta*, 1967, **23A**, 2699–2700. [78, 91]
46. Crisler, R. O. and Brubaker, I. M., Technique for establishing baseline in IR spectrometry, *Appl. Spectros.*, 1967, **21**, 126–128. [193]
47. Crocket, D. S. and Haendler, H. M., Halocarbon oil as a mulling medium for IR spectra, *Analyt. Chem.*, 1959, **31**, 626–627. [24]
48. Cross, A. D. and Jones, R. A., *An Introduction to Practical Infrared Spectroscopy* (3rd edn, Butterworths, 1969); see p. 21. [27, 28, 29, 36]
49. Cross, L. H., Haw, J. and Shields, D. J., Retrieval of IR data, *Proc. 4th Conf. Mol. Spectrosc.*, 1968, 189–205. [165, 187]
50. Curry, A. S., *et al.*, Micro IR spectroscopy of gas chromatographic fractions, *J. Chromatog.*, 1968, **38**, 200–208. [78, 91]
51. Curry, A. S., Read, J. F. and Brown, C., A simple IR spectrum retrieval system, *J. Pharm. Pharmac.*, 1969, **21**, 224–231. [168, 174, 181, 209, 211ff.]
52. Curry, D. R., Information retrieval in the analytical laboratory: A review, *Analyst., Lond.*, 1963, **88**, 829–834. [171, 174]
53. Daniels, N. W. R., GC–IR microanalysis, *Column* (W. G. Pye Gas Chromatography Bulletin), 1967, **2** (1), 2–5. [76]
54. Dannenberg, H., Forbes, J. W. and Jones, A. L., IR spectroscopy of surface coatings in reflected light, *Analyt. Chem.*, 1960, **32**, 365–370. [35]
55. de Faubert Maunder, M. J., Egan, H. and Roburn, J., Some practical aspects of the determination of chlorinated pesticides by electron-capture gas chromatography, *Analyst, Lond.*, 1964, **89**, 157–167. [75]
56. de Faubert Maunder, M. J., Report on the analysis of one tablet of STP, *Confidential Newsletter Ass. Publ. Analysts*, 1967, 52–55. See also Ref. 151. [100]
57. de Faubert Maunder, M. J., Simple chromatography of cannabis constituents, *J. Pharm. Pharmac.*, 1969, **21**, 334–335. [82]
58. Degen, I. A., Detection of methoxyl groups by IR, *Appl. Spectros.*, 1968, **22**, 164. [109]
59. de Klein, W. J., IR spectra of compounds separated by TLC using KBr micro pellet technique, *Analyt. Chem.*, 1969, **41**, 667–668. [61, 82]
60. de Klein, W. J. and Ulbert, K., A simple micropelleting technique, *Analyt. Chem.*, 1969, **41**, 682. [21]
61. Determination of sodium propionate content—Appendix H, British Standard 4307: 1968; *Specification for Calcium Propionate and Sodium Propionate for Use in Foodstuffs* (British Standards Institution, London, 1968). [88]
62. Devaney, R. G. and Thompson, A. L., An improved technique for preparing samples or organic solids for IR analysis, *Appl. Spectros.*, 1958, **12**, 154–155. [32]

63. *The Documentation of Molecular Spectroscopy* (DMS) (Butterworths, London; Verlag Chemie, Weinheim-Bergstrasse). [164]
64. DOLE, V. P., KIM, W. K. and EGLITIS, I., Detection of narcotic drugs, tranquillisers, amphetamines, and barbiturates in urine, *J. Amer. Med. Ass.*, 1966, **198**, 349–352. [87, 92]
65. DOLE, V. P., KIM, W. K. and EGLITIS, I., Extraction of narcotic drugs, tranquillisers, and barbiturates by cation-exchange paper, and detection on a thin-layer chromatogram by a series of reagents, *Psychopharmacol. Bull.*, 1966, **3**, 45–48. [87]
66. DOLINSKY, M. J., A technique for IR analysis of solids insoluble in non polar solvents, *J.A.O.A.C.*, 1951, **34**, 748–763. [32]
67. DOLINSKY, M. J., WENNINGER, J. A. and SCHMIDT, W. E., Improved suspension technique for quantitative analysis of insoluble solids, *Analyt. Chem.*, 1963, **35**, 931–933. [32]
68. DOYLE, T. D. and LEVINE, J., Application of ion-pair extraction to partition chromatographic separation of pharmaceutical amines, *Analyt. Chem.*, 1967, **39**, 1282–1287. [89]
69. DUYCKAERTS, G., The IR analysis of solid substances: A review, *Analyst, Lond.*, 1959, **84**, 201–214. [52, 64, 71]
70. EDWARDS, G. J., A rectangular KBr pellet die and holder, *Appl. Spectros.*, 1964, **18**, 94–95. [15]
71. ERLEY, D. S., Fast searching system for the ASTM IR data file, *Analyt. Chem.*, 1968, **40**, 894–898. [165]
72. Extrocells are available from Research and Industrial Instruments Company, Worsley Bridge Road, London, SE 26. [51, 76]
73. FAHRENFORT, J., Attenuated total reflection. A new principle for the production of useful IR reflection spectra of organic compounds, *Spectrochim. Acta*, 1961, **17**, 698–709. [35]
74. FALES, H. M., *et al.*, Milligram-scale preparative gas chromatography of steroids and alkaloids, *Anal. Biochem.*, 1962, **4**, 296–305. [76, 77, 79]
75. FARMER, V. C., Effects of grinding during the preparation of alkali-halide disks on the IR spectra of hydroxylic compounds, *Spectrochim. Acta*, 1957, **8**, 374–389. [75]
76. FEAIRHELLER, W. R. and DU FOUR, H., Polishing jig for IR cell windows, *Appl. Spectros.*, 1967, **21**, 45. [50]
77. FINKEL'SHTEIN, A. I., *et al.*, Simple method for preparing plates of KBr for the IR spectroscopy of solids, *Optics Spectros.*, 1959, **6**, 415–417. [9]
78. FIORI, A. and MARIGO, M., Some chromatographic problems in the chemical toxicology of alkaloids: A simple solution, *Zacchia*, 1964, **39**, 317–331. [81]
79. FURCHGOTT, R. F., ROSENKRANTZ, H. and SHORR, E., IR absorption spectra of steroids; I, Androgens and related steroids, *J. Biol. Chem.*, 1946, **163**, 375–386. [36]
80. GARNER, H. R. and PACKER, H., New technique for the preparation of KBr pellets from micro samples, *Appl. Spectros.*, 1968, **22**, 122–123. [85]
81. GIOVANELLI, R. G., Chemical analysis by diffuse reflexion spectrophotometry, *Nature, Lond.*, 1957, **179**, 621–622. [35]
82. GLASSNER, S., IR spectra of small samples without beam condenser, *Appl. Spectros.*, 1962, **16**, 112–113. [50]

83. GOENECHIA, S., Preparative TLC pretreatment of soporific mixtures prior to IR identification, *Fresenius' Z. Anal. Chem.*, 1967, **225**, 30–36. [81]

84. GROSTIC, M. F. and BRANSON, G. E., Use of resolidified melts in IR spectroscopy: Application to polymorphism and solvation studies, *Appl. Spectros.*, 1961, **15**, 157–159. [70]

85. HAJRA, A. K. and RADIN, N. S., Collection of GLC effluents, *J. Lipid Research*, 1962, **3**, 131–134. [77]

86. HALES, J. S. and KYNASTON, W., The preparation of pressed disks of purified KCl containing solid samples for IR spectrometry, *Analyst, Lond.*, 1954, **79**, 702–706. [6, 8]

87. HAYDEN, A. L. and SAMMUL, O. R., IR analysis of pharmaceuticals; I, Application of the KBr disk technique to some steroids, alkaloids, barbiturates and other drugs, *J. Am. Pharm. Assoc.*, 1960, **49**, 489–496. [54]

88. HAYDEN, A. L., *et al.*, IR, UV, and visible absorption spectra of some USP and NF reference standards and their derivatives, *J.A.O.A.C.*, 1962, **45**, 797. [167]

89. HAYDEN, A. L., BRANNON, W. L. and YACIW, C. A., Drugs: IR spectra of some compounds of pharmaceutical interest, *J.A.O.A.C.*, 1966, **49**, 1109–1153. See spectrum no. 553, p. 1112. [167]

90. HOFMANN, B. R. and ELLIS, G. H., Easy method of salt formation as aid to structure elucidation in IR spectrometry, *Analyt. Chem.*, 1967, **39**, 406–407. [79, 198]

91. HUBLEY, C. E. and LEVI, L., Physical methods for the identification of narcotics; Part IVA, The infrared spectroscopic method, *Bulletin on Narcotics*, 1955, **7**, 20–41; see p. 29. [28]

92. HUO-PING, P. and EDWARDS, G. J., A new technique for preparing films on cell windows for IR absorption spectroscopy, *Appl. Spectros.*, 1963, **17**, 74–75. [39]

93. *Infrared Correlation Chart* (Barnes Engineering Company/Instrument Division: Available from Field Instruments Co. Ltd., Tetrapak House, Orchard Road, Richmond, Surrey). [196]

94. INGEBRIGSTON, D. N. and SMITH, A. L., IR analysis of solids by KBr pellet technique, *Analyt. Chem.*, 1954, **26**, 1765–1768. [9, 12]

95. JAFFE, J. H. and KIRKPATRICK, D., The use of ion-exchange resin impregnated paper in the detection of opiate alkaloids, amphetamines, phenothiazines and barbiturates in urine, *Psychopharmacol. Bull.*, 1966, **3**, 49–52. [87, 92]

96. JOHNSON, C. A., Book review, *Pharm. J.*, 1969, April 26, 455–456. [167]

97. JONES, R. N., *et al.*, The use of indene for the calibration of small IR spectrometers, *Spectrochim. Acta*, 1961, **17**, 77–81. [197]

98. KENDALL, R. F., Methods for rapid transfer of GLC fractions into IR cavity cells, *Appl. Spectros.*, 1967, **21**, 31–32. [76]

99. KING, G. W., BLANTON, E. H. and FRAWLEY, J., Spectroscopy from the point of view of communication theory; IV, Automatic recording of IR spectra on punched cards, *J. Opt. Soc. Am.*, 1954, **44**, 397–402. [165]

100. KIRKLAND, J. J., Quantitative application of KBr disk technique in IR spectroscopy, *Analyt. Chem.*, 1955, **27**, 1537–1541. [9, 11]

101. KOENIG, J. L., Effect of nonuniform distribution of absorbing material on the quantitative measurement of IR band intensities, *Analyt. Chem.*, 1964, **36**, 1045–1046. [17, 29]

102. KOPFF, R., RUEFF, R. and BENOÎT, H., Some improvements in the technique using potassium bromide as support for IR spectroscopic samples, *Bull. Soc. Chim. France*, 1960, 1694–1695. [8]

103. KRAMER, A., Convenient method for encoding IR spectra linear in wavenumber using Sadtler notation, *Appl. Spectros.*, 1967, **21**, 184–185. [197]

104. LAMB, D. J. and BOPE, F. W., Preparation and properties of 8-chlorotheophylline salts of some nitrogen bases, *J. Am. Pharm. Assoc., Sci. Ed.*, 1954, **45**, 178–181. [91]

105. LAUNER, P. J., Tracking down spurious bands in IR analysis, published in *Laboratory Methods in Infrared Spectroscopy*, R. G. J. MILLER (ed.) (Sadtler Research Laboratories Inc., 1965), pp. 13–17. [58]

106. LEGGON, H. W., A KBr powder trap for gas chromatographs for obtaining IR spectra, *Analyt. Chem.*, 1961, **33**, 1295–1296. [78]

107. LEVI, L. and FARMILO, C. G., Reaction of morphine with potassium-mercuric iodide, *Analyt. Chem.*, 1954, **26**, 1040–1045. [90]

108. LEVI, L., HUBLEY, C. E. and HINGE, R. A., Physical methods for the identification of narcotics, Part IVB: Infrared spectra of narcotics and related alkaloids, *Bulletin on Narcotics*, 1955, **7**, 42–84; mull spectra 48–69. [28]

109. LEVI, L. and HUBLEY, C. E., Detection and identification of clinically important barbiturates, *Analyt. Chem.*, 1956, **28**, 1591–1605. [92, 107]

110. LEVI, L., The morphine-Marmé complex, *Analyt. Chem.*, 1957, **29**, 470–474. [90]

111. LEVINE, J., Analysis of organic bases by salt partition, *J. Pharm. Sci.*, 1965, **54**, 485–488. [89]

112. LEYSEN, R. and VAN RYSSLBERGE, J., Preclassification system for the Wyandotte-ASTM punched cards, indexing spectral absorption data, *Appl. Spectros.*, 1965, **19**, 72–74. [165]

113. LOHR, L. J. and KAIER, R. J., Preparation of micro nujol mulls for IR analysis, *Analyt. Chem.*, 1960, **32**, 301. [25]

114. LOW, M. J. D. and FREEMAN, S. K., Measurement of IR spectra of GLC fractions using multiple-scan interference spectrometry, *Analyt. Chem.*, 1967, **39**, 194–198. [80]

115. LOW, M. J. D., Applications of multiple scan interferometry to the measurement of IR spectra, *Appl. Spectros.*, 1968, **22**, 463–471. [80]

116. LOW, M. J. D. and ABRAMS, L., Anomalies in IR transmission spectra caused by the self-emission of translucent samples, *Appl. Spectros.*, 1966, **20**, 416–417. [73]

117. MAEHLEY, A. C., A micro technique for identifying barbiturates in forensic chemistry, *Analyst, Lond.*, 1962, **87**, 116–120. [91]

118. 'Manisort' coincidence cards. Available from George Anson, Ref. 8. [182]

119. MANNING, J. J. and O'BRIEN, K. P., The analysis of barbiturates by the pressed bromides sampling method and IR spectrophotometry, *Bull. Narcot.*, 1958, **10**, 25–34. [107]

120. MANNO, R. P., PARASKEVOPOULOS, N. and MATSUGUMA, H. J., A variation of the compensation technique in IR spectrophotometry, *Appl. Spectros.*, 1959, **13**, 57–59. [54]

121. MARION, L., RAMSAY, D. A. and JONES, R. N., The IR absorption spectra of alkaloids, *J.A.C.S.*, 1951, **73**, 305–308. [109]
122. MARTIN, A. E., Identification of organic compounds from IR spectra, *Nature, Lond.*, 1952, **170**, 20–22. [165]
123. 'MARTINDALE', *Extra Pharmacopoeia*, R. G. TODD (ed.) (The Pharmaceutical Press, London, 25th edn, 1967). [200]
124. MATTA, G., SILVA, M. J. and LOPES, M. M. S., Sodium tetraphenylborate as a reagent for identification and assay of organic bases, *Rev. Port. Farm.*, 1965, **15**, 341–357. [90]
125. MAY, L., *Spectroscopic Tricks* (Adam Hilger Ltd, London, and Plenum Press, New York, 1968). [4]
126. MCARDLE, C. and SKEW, E. A., A scheme for rapid tablet identification, *Lancet*, 1961, **2**, 924. Modified cards now used by the United Birmingham Hospitals and other hospitals overseas. Cards available from Copeland-Chatterson Co. Ltd, 1 Watling Street, London, E.C.4. [179, 180]
127. MCARDLE, C. and SKEW, E. A., Solid dosage forms—An aid to their rapid identification, *Symposium: Identification of Drugs and Poisons* (Pharmaceutical Press, London, 1965), pp. 1–8. [178]
128. MCCOY, R. N. and FIEBIG, E. C., Technique for obtaining IR spectra of microgram amounts of compounds separated by TLC, *Analyt. Chem.*, 1965, **37**, 593–595. [84]
129. MCDEVITT, N. T. and BAUN, W. L., Contamination of KBr pellets by plastic mixing vials, *Appl. Spectros.*, 1960, **14**, 135–136. [11, 62]
130. MCGAUGHRAN, W. R., Filing KBr pellets, *Appl. Spectros.*, 1956, **10**, 64. [21]
131. *Medinex*, PULLOM, E. N. (ed.) (Haymarket Press Ltd, London, 1967), now superseded by *MIMS Annual Compendium*, published with the April issue of Ref. 146. [202]
132. MELOCHE, V. W. and KALBUS, G. E., Anomalies in the IR spectra of inorganic compounds prepared by the KBr pellet technique, *J. Inorg. Nucl. Chem.*, 1958, **6**, 104–111. [70]
133. STECHER, P. G. (ed.), *The Merck Index* (Merck & Co., Inc., Rahway, N.J., USA, 8th edn, 1968). [199]
134. MESLEY, R. J., The IR spectra of steroids in the solid state, *Spectrochim. Acta*, 1966, **22**, 889–917. [107]
135. MESLEY, R. J., Spectra-structure correlations in polymorphic solids; II, 5,5-disubstituted barbituric acids, *Spectrochim, Acta*, 1970, **26A**, 1427–1448. [107]
136. MESLEY, R. J. and CLEMENTS, R. L., IR identification of barbiturates with particular reference to the occurrence of polymorphism, *J. Pharm. Pharmac.*, 1968, **20**, 341–347. [91]
137. MESLEY, R. J. and EVANS, W. H., IR identification of lysergide (LSD), *J. Pharm. Pharmac.*, 1969, **21**, 713–720. [109]
138. MESLEY, R. J. and EVANS, W. H., IR identification of some hallucinogenic derivatives of tryptamine and amphetamine, *J. Pharm. Pharmac.*, 1970, **22**, 321–332. [108]
139. MILKEY, R. G., KBr method of IR sampling, *Analyt. Chem.*, 1958, **30**, 1931–1933. [9]
140. MILLER, R. G. J., *Laboratory Methods in Infrared Spectroscopy* (Plenum Press, New York, 1964). [4]

141. MILLETT, M. A., MOORE, W. E. and SAEMAN, J. F., Technique for quantitative TLC, *Analyt. Chem.*, 1964, **36**, 491–494. [84]

142. MILLS, A. L., IR identification of microgram quantities of heroin hydrochloride, *Analyt. Chem.*, 1963, **35**, 416. [36]

143. MITZNER, B. M., Gelatine capsules for potassium bromide IR technique, *Analyt. Chem.*, 1956, **28**, 1801. [10]

144. MITZNER, B. M., A simple sample measuring device for use with the KBr technique as applied to IR spectroscopy, *Appl. Spectros.*, 1956, **10**, 75–76. [10]

145. MOLNAR, W. S. and YARBOROUGH, V. A., Polyethylene capillary tubes as micro liquid cells for IR spectroscopy, *Appl. Spectros.*, 1958, **12**, 143. [76]

146. 'MINS', *Monthly Index of Medical Specialities*, WILSON, F. J. (ed.) (Haymarket Press Ltd, London). [202]

147. MORGAN, H. W., A levelling device for alkali halide pressed disk preparation, *Appl. Spectros.*, 1959, **13**, 48. [18]

148. *Multilingual List of Narcotic Drugs under International Control* (Document E/CN. 7/513. Published by United Nations, New York, 1968, Sales No. E/F/S/R.69.XI.1). [200]

149. NICHOLLS, S. F., The preparation of small rock salt plates for microsampling in IR spectrophotometry, *Analyst., London*, 1961, **86**, 664–666. [50]

150. PALMER, H. K., Circular correlation charts for the assignment of bands in IR spectra, *Appl. Spectros.*, 1964, **18**, 191–193. [196]

151. PHILLIPS, G. F. and MESLEY, R. J., Examination of the hallucinogen 2,5-dimethoxy-4-methylamphetamine, *J. Pharm. Pharmac.*, 1969, **21**, 9–17. [109]

152. *Physicians Desk Reference to Pharmaceutical Specialities and Biologicals* (published annually by Medical Economics Inc., Oradell, N.J., USA). [202]

153. PRIEST, L. (ed.), *Poisons and TSA Guide* (The Pharmaceutical Press, London, 9th edn, 1968). [200]

154. POLCHLOPEK, S. E. and ROBERTSON, H. J., Minimizing the polystyrene contamination in Wig-L-Bug grinding, *Appl. Spectros.*, 1962, **16**, 112. [12]

155. POTTS, W. J., IR characterization of side-chain substitution of monoalkyl benzenes, *Analyt. Chem.*, 1955, **27**, 1027–1030. [109]

156. PRICE, G. D., *et al.*, Microcell for obtaining normal contrast IR solution spectra at the 5 μg level, *Analyt. Chem.*, 1967, **39**, 138–140. [50]

157. PRICE, W. H. and MAURER, R. H., Preparation of KBr pellets of unstable materials, *Appl. Spectros.*, 1963, **17**, 106–107. [19]

158. RAPPAPORT, G., Specular IR spectral reflectance for the analysis of organic materials, *Phys. Rev.*, 1948, **74**, 115. [35]

159. Regina v. Graham, Court of Criminal Appeal, 24th January, 1969: D.D. (No. 2) Regs., 1964 Reg. 3: *Times Law Report*, 28th January, 1969. [56]

160. Regina v. Worsell, Court of Criminal Appeal, 13th December, 1968: D.D. (No. 2) Regs., 1964, Reg. 9: *Times Law Report*, 14th December, 1968. [55]

161. RESNIK, F. E., *et al.*, IR microtechniques for identification of carbohydrates and other organic compounds, *Analyt. Chem.*, 1957, **29**, 1874–1877. [52]

162. RICE, D. D., Direct transfer technique for preparing micropellets from TLC chromatograms for IR identification, *Analyt. Chem.*, 1967, **39**, 1906–1907. [84]

163. RICH, N. W. and CHATTEN, L. G., Identification and differentiation of organic medicinal agents; I, Local anaesthetics, *J. Pharm. Sci.*, 1965, **54**, 995–1002. [91]

164. RIDGWAY WATT, P., A mechanical auxilliary recorder for an IR spectrometer, *Chemy. Ind.*, 1959, **78**, 44–45. [177]

165. ROBEY, R. F. (Secretary to the Committee), Ten years of aid to applied absorption spectroscopy by ASTM Committee, E–13, *Appl. Spectros.*, 1960, **14**, 103–106. [164]

166. ROBINSON, D. E., Quantitative analysis with IR spectrophotometers: differential analysis, *Analyt. Chem.*, 1952, **24**, 619–622. [53]

167. ROBINSON, T. S. and PRICE, W. C., The determination of IR absorption spectra from reflection measurements, *Proc. Phys. Soc.* (*London*), 1953, **66B**, 969–974. [35]

168. *The Sadtler Catalogue of Standard Spectra* (Sadtler Research Laboratories, 3314–20, Spring Garden Street, Philadelphia, Penna. 19104, USA). [162]

169. SANDS, J. D. and TURNER, G. S., New development in solid phase IR spectroscopy, *Analyt. Chem.*, 1952, **24**, 791–793. [33]

170. SAUMAGNE, P. and JOSIEN, M. L., *Perkin-Elmer Instr. News*, 1959, **10** (4), 12. [7]

171. SAVITZKY, A., Data processing in analytical chemistry, *Analyt. Chem.*, 1961, **33** (13), 25A-46A. Correction: *ibid.*, 1962, **34** (1), 32A. [165]

172. SCHIEDT, U., German technique for pressed disk pressed KBr IR sampling, *Appl. Spectros.*, 1953, **7**, 75–84. [6, 12]

173. SCHIEDT, U. and REINWEIN, H., The IR spectroscopy of amino acids; I, A new preparation technique for the IR spectroscopy of amino acids and other polar compounds, *Z. Naturf.*, 1952, **7b**, 270–277. [5, 12]

174. SCHLICHTER, N. E. and WALLACE, E., 'Peek-a-boo' data retrieval in IR spectroscopy, *Appl. Spectros.*, 1963, **17**, 98–101. [168, 182]

175. SESHADRI, K. S. and JONES, R. N., The shapes and intensities of IR absorption bands—A Review, *Spectrochim. Acta*, 1963, **19**, 1013–1085. [193]

176. SINSHEIMER, J. E. and SMITH, E., Identification of sympathomimetic amines as tetraphenylborates, *J. Pharm. Sci.*, 1963, **52**, 1080–1085. [90]

177. SINSHEIMER, J. E. and KEUHNELIAN, A. M., Near-IR spectroscopy of amine salts, *J. Pharm. Sci.*, 1966, **55**, 1240–1244. [109]

178. SPITTLER, T. M. and JASELSKIS, B., Preparation and use of powdered silver chloride as IR matrix, *Appl. Spectros.*, 1966, **20**, 251. [7, 9]

179. STANFIELD, J. E., SHEPPARD, D. E. and HARRISON, H.S., Sample handling, published in *Laboratory methods in IR spectroscopy*, R. G. J. MILLER (ed.) (Heyden & Son Ltd, London, 1965); see pp. 45–46. [27, 29]

180. STERLING, K. J., Preparation of KBr disks for IR microanalysis by use of a half inch die, *Analyt. Chem.*, 1966, **38**, 1804. [49]

181. STEVENS, H. M., The breaking of emulsions in toxicological extractions, *J. For. Sci. Soc.*, 1968, **8**, 66. [86]

182. STEWART, J. E., Sampling technique for IR spectroscopy of solids, *Analyt. Chem.*, 1959, **31**, 1287. [34]

183. STIMSON, M. M. and O'DONNELL, M. J., The IR and UV absorption spectra of cytosine and isocytosine in the solid state, *J.A.C.S.*, 19[illegible], **74**, 1805–1808. [5]

184. SUNSHINE, I. and GERBER, S. R., *Spectrophotometric Analysis of Drugs including an Atlas of Spectra* (Charles C. Thomas, Ltd, Springfield, Illinois, USA, 1963); see the spectrum of dextroamphetamine sulphate in a KBr pellet, p. 134. [167]

185. SZONYI, C. and CRASKE, J. D., A simple micromull technique for obtaining IR spectra, *Analyt. Chem.*, 1962, **34**, 448. [26, 32, 95]

186. THOMAS, I. C., *A New Chemical Structure Code for Data Storage and Retrieval in Molecular Spectroscopy* (Heyden & Sons, London, 1968). [169]

187. THOMAS, P. J. and DWYER, J. L., Collection of gas-chromatographic effluents for IR spectral analysis, *J. Chromatog.*, 1964, **13**, 366–371. [78]

188. THOMPSON, W. E., *et al.*, Identification of primary, secondary and tertiary pharmaceutical amines by the IR spectra of their salts. *J. Pharm. Sci.*, 1965, **54**, 1819–1821. [109]

189. THOMPSON, W. H., The documentation of molecular spectra, *J. Chem. Soc.*, 1955, 4501–4509. [64]

190. THOMPSON, W. K., An IR absorption study of the state of adsorbed water molecules and the effect of sodium ions on KBr pressed disks, and on dipotassium hydrogen phosphate dispersed in KBr pressed disks, *J. Chem. Soc.*, 1964, 3658–3663. [72]

191. TOLK, A., The alkali halide disk technique in IR spectrometry: Anomalous behaviour of some samples dispersed in alkali halide disks, *Spectrochim. Acta*, 1961, **17**, 511–522. [70, 71]

192. TROXLER, F. and HOFMAN, A., Oxidation of lysergic acid- 2,3 derivatives, *Helv. Chim. Acta*, 1959, **42**, 793–802 (comparisons). [109]

193. UMBERGER, C. J. and ADAMS, C., Identification of malonyl urea derivatives IR absorption in toxicological analysis, *Analyt. Chem.*, 1952, **24**, 1309–1322 (solution spectra). [107]

194. VANDENBELT, J. M., SCOTT, R. B. and SCHOEB, E. J., Pantographic reduction of IR spectrograms, *Appl. Spectros.*, 1954, **8**, 88. [178]

195. VON DIETRICH, H., An expedient method for preconditioning of KBr as an embedding medium for IR spectroscopy, *Z. Naturf.*, 1956, **11b**, 175–176. [9]

196. WARREN, R. J., *et al.*, Spectra-structure correlations of phenothiazines by IR, UV, and NMR spectra, *J. Pharm. Sci.*, 1966, **55**, 144–150. [108]

197. WARREN, R. J., *et al.*, Pharmaceutical applications of internal reflectance spectroscopy, *Microchem. J.*, 1967, **12**, 555–567. [35]

198. WASHBURN, W. H., Specific procedure for running differential surveys by IR, *Appl. Spectros.*, 1956, **10**, 46. [53]

199. WASHBURN, W. H. and MAHONEY, M. J., Use of substitute standards in IR differential spectrophotometry, *Analyt. Chem.*, 1958, **30**, 1053–1055. [54]

200. WAYLAND, L. and WEISS, P. J., IR spectra of some antibiotics of interest, *J.A.O.A.C.*, 1965, **48**, 965–972. [108]

201. WHITE, R. G., *Handbook of Industrial IR Analysis* (Plenum Press, New York, 1964). [4, 204]

202. Wick-Sticks, a typical advert., *Analyt. Chem.*, 1967, **39** (13), **91A**, available from Harshaw Chemical Co. [85]

203. WIDMARK, J. and WIDMARK, G., Quantitative gas chromatography. Quantitative recovery and reinjection of a sample, *Acta Chem. Scand.*, 1962, **16**, 575–582. [77, 78]

204. WILKS, P. A., Plate for combined attenuated total reflectance and thin-layer chromatographic analysis, *U.S. Patent*, 1966, 3 279 307. [86]

205. WILKS, P. A. and HIRSCHFELD, T., Internal reflection spectroscopy, *Appl. Spectr. Rev.*, 1967, **1**, 99–130. [35]

206. WILKS, P. A., Internal reflection spectroscopy. I: Effect of angle of incidence change, *Appl. Spectros.*, 1968, **22**, 782–784; II: Quantitative analysis aspects, *ibid.*, 1969, **23**, 63–66. [35, 51]

207. Wilks MIR–15, GC–IR analyser, available from Techmation Ltd, 58 Edgware Road, Edgware, Middlesex. [76]

208. Wilks Model 41 Vapour phase GC–IR Analyser, available from Techmation Ltd, 58 Edgware Way, Edgware, Middlesex. [80]

209. YUNG, D. K. and PERNAROWSKI, M., Identification and differentiation of some phenothiazine-type tranquillisers, *J. Pharm. Sci.*, 1963, **52**, 365–370. [90, 91]

210. ZINK, T. H., Novel cataloguing system, *Appl. Spectros.*, 1961, **15**, 22–23. [168]

211. The A.O.A.C. collection contains typical sulphonamide spectra: Indexes—1962, **45**, 898–900; 1964, **47**, 918–924; 1966, **49**, 1150–1153. [107, 211ff]

INDEX